BLUEBIRD
SEASONS

BLUEBIRD SEASONS

Witnessing Climate Change
in My Piece of the Wild

MARY TAYLOR YOUNG

Copyright © 2023 by Mary Taylor Young
Illustrations copyright © 2023 by Richard K. Young
All rights reserved
Published by Chicago Review Press Incorporated
814 North Franklin Street
Chicago, Illinois 60610
ISBN 978-1-64160-813-8

Library of Congress Control Number: 2022950091

Cover images: bluebird, InnovativeImages/Shutterstock; abstract painting,
Sweet Art / Shutterstock; mesa and sky, author's collection
Cover design: Jonathan Hahn
Typesetting: Nord Compo

Printed in the United States of America
5 4 3 2 1

This book is dedicated to my family—two legged and four legged—who walked this path with me and who have loved our piece of the wild with equal passion. Rick, Olivia, Cody, Margo, Jasper, Rosie, Jesse, and Katie.

Only that day dawns to which we are awake.
There is more day to dawn. The sun is but a morning star.

—Henry David Thoreau

When we see land as a community to which we belong,
we may begin to use it with love and respect.

—Aldo Leopold, *A Sand County Almanac*, 1949

CONTENTS

Preface
CALL ME BY MY NAME

I began work on a version of this book more than ten years ago, intending a lyrical nature memoir about my family's years among the forests, mountains, and meadows of southern Colorado. I planned to base that early incarnation—*A Bluebird Season*—on the nature journal we have kept since autumn 1995, when we first purchased our land in the foothills of the Sangre de Cristo Range in the southern Rocky Mountains. Our experiences keeping a trail of bluebird nest boxes would be the centerpiece, a way to reveal the joys and spiritual renewal we found in nature from intimately watching one piece of land over many seasons and many years.

Over time, though, my sweet and simple story grew more serious. We had observed and delighted in the natural rhythms of the land for two decades, renewed and reassured by them. But there were more profound changes already underway on the earth and in the skies. I realized our journal, when looked at over the long term, reflected many of these changes, not just seasonal or cyclical variations, like dry years followed by rainy ones, but also deeper and long-lasting differences.

The more I thought about the alterations we were seeing, the more evident it became we were witnesses to the growing effects of climate change. This was serious stuff—long-term drought, increased wildfires, die-off of piñon forests, extremes of weather, the gradual disappearance of once-common birds and arrival of more south-ranging species.

I could no longer proceed with my gentle memoir of our life among the bluebirds with any authenticity or integrity. As often happens with writing, the story had emerged that needed to be told and it was shouting at me louder and louder—*Call me by my name!*

With new eyes, I studied our journal from twenty-five years on the land. Embedded in the entries was the evidence. The journal was more than just a nice record of sightings, of species, of weather and happenings. It was a chronicle of climate change. And it was happening not far away, in an Indonesian rain forest or Antarctic ice sheet, but much closer to home. Here, in our piece of the wild.

And so the simple story of *A Bluebird Season* became *Bluebird Seasons*, a testament to a world that had altered, and was still altering. The stories of finding joy and renewal in nature are still the fabric of my tale, but they are tempered by the warning of environmental upheaval that is no longer on the horizon, but right here and right now.

Climate change is part of everyone's daily life. It's evident in all our backyards, in everyone's piece of the wild. In winter days much warmer than they should be. In record-high temperatures, day after week after month. In spring flowers coming up in January. In birds that should have migrated a month ago or new species showing up at the feeder and staying to nest. In the parched grass and trees and landscaping from rains that haven't come or in unprecedented rains and devastating floods. In the skeletal trees long hidden below the waters of the local reservoir, now standing bare and exposed, or in shorelines lost to rising seas. I hope my story will prompt readers to look around their own backyards, their own pieces of the wild, then call what is happening by its name.

This book is the story of past bluebird seasons. The tale of future seasons waits to be written. The thing about seasons is that their ultimate dynamic is change—birth, growth, death, renewal. There are a variety of possible endings to the story, different paths we as a global village can choose to take. We can keep the seasons turning past loss to renewal.

In that lies our hope.

ACKNOWLEDGMENTS

Thanks with all my heart to Rick and Olivia; we have walked this entire journey together at our special place. Without you both, I would not have made this journey or written this book. The three of us shared a large piece of our lives at the cabin—our piece of the wild. Thanks for helping me with memories of the last twenty-five years there. And thank you Rick for creating the space for me to devote an intense seven months to writing this book and for reading and offering feedback on every chapter as well as sharing your wonderful illustrations.

Thank you to all the friends who pitched in and helped us build the cabin, shared community and stories, and broke bread with us around the campfire—Paul Gray, John White, Bill Cole, Deb Long, Becky Jones, Clark Wilson, Carolyn Sutton, Sally Taylor, Deanna Vaughn, Mike Adam, Doug Sacarto, Lisa Hutchins, Suzanne and F. B. Becquet, Mel and Bud, and John Curtis. I have likely forgotten someone, but know that I appreciate and thank you for your contribution.

Thank you to Art Trujillo for sharing his stories, and Robin and Kevin Shishakowsky for sharing their experience in the drilling industry and for being terrific advocates for our community.

Mark D. Mitchell, research director, Paleocultural Research Group in Arvada, Colorado, shared his time and expertise with me in a fascinating discussion of the Sopris Phase culture, Rio Grande pottery, stone tools, and other archaeological details. Thanks so much. We could have talked for hours!

The Louden-Henritze Archaeology Museum at Trinidad State College was helpful in background on the geology and archaeology of the area, particularly early human cultures and the asteroid impact.

Trinidad Lake State Park's interpretive site showcasing its piece of the K-Pg Boundary gets a thanks for being literally older than dirt and so cool—a rare spot where you can touch the land before time.

Many thanks to my second daughter, Anna Rose Lowenthal, who has been visiting the cabin since she was three years old, for sharing her vast and passionate knowledge about sustainable fashion. I still see you and Olivia as five-year-olds, hulaing on the deck as the hummingbirds buzzed so noisily around you they drowned out your *A-LO-ha*s.

And again to my Olivia Pearl, the Animal Girl—who has migrated on from the cabin but keeps it always in her heart—for sharing the passion and hope for the future of herself and her generation.

To Jerry Pohlen of Chicago Review Press, thank you for understanding my vision for this book and the critical role of story, not just science, in the literature of climate change. Thanks also to all the crew there for making my words into a wonderful book and getting my story and my message fledged and out of the nest for the world to read.

INTRODUCTION

This is not a matter of Chicken Little telling us the sky is
falling. The scientific evidence . . . is telling us we have a
problem, a serious problem.
—Senator John H. Chafee (R-RI), June 1986, remarks
   from the US Senate Committee on the Environment
   and Public Works hearings on "Ozone Depletion,
   the Greenhouse Effect, and Climate Change"

We have broken our covenant with the natural world. We have
failed as stewards of the place that sustains us. That is the real
cause of climate change.

Wealthy countries bear the brunt of the blame, but given a chance,
people everywhere will exploit Earth's resources if it means a better life.
It's only human. We are driven biologically to survive and succeed the
best way we can. And we're not very good at balancing that drive with
maintaining a healthy, sustainable ecosystem. We have difficulty seeing
the big picture of how we're hurting the forest when we're busy with
the trees. We don't notice that we are sitting on the tree limb we are
sawing off.

   Twenty-five years ago, I wrote a book called *Land of Grass and Sky:
A Naturalist's Prairie Journey*, in which I said the environmental catas-
trophe of the 1930s known as the Dust Bowl happened not because
of a drought—there had been worse droughts before and since—but

because the people farming the Great Plains broke their covenant with the land. Instead of being stewards who listened to the land and worked it sustainably within its rules, they became plunderers.

The same is true today, but our broken covenant is with the entire planet. By heating the atmosphere through the release of massive amounts of carbon, we're causing catastrophic changes in the natural systems of Earth that have a real risk of making it uninhabitable for ourselves and for many of the life-forms—plants, animals, microorganisms—with which we share this spinning globe.

This thing we have wrought, this thing called climate change, is the largest single threat to the survival of the world as we know it.

———————

I have devoted my professional life to environmental conservation. Over thirty-plus years as a nature and wildlife writer, I've tried to touch minds with intriguing information and, more important, to touch hearts by sharing the delight and wonder I find in nature. I've written with humor and emotion to persuade people to support environmental conservation, to put their voices and their donations and their votes to preserving the natural world we all share.

The threats to the environment I've been writing about for decades—species decline, habitat loss, pollution, pesticides, urban and suburban expansion—haven't gone away. But warnings of a greater, even more challenging threat have gotten louder and louder until they are deafening.

Earth's temperature is rising, scientists warn us. At first this phenomenon was called *global warming*, but that term confused people. If things were warmer, why were we having unprecedented snowstorms and weird events like bomb cyclone blizzards? *Climate change* was a better term.

As early as 1986, Congress began investigating this threat. Through the 1990s, climatologists and scientists of many disciplines increasingly warned us that climate change is real and that it is serious, folks. But we weren't really listening. Global warming remained the subject of jokes on late-night television. We all kept buying gas-powered cars, cranking the heat or the air conditioning to keep ourselves comfy, and listening to naysayers with self-serving agendas.

Climate change didn't really break into the broad public consciousness until Vice President Al Gore's 2006 film *An Inconvenient Truth* won an Academy Award for Best Documentary Feature. (It also won Best Original Song for "I Need to Wake Up" by Melissa Etheridge.) I bought several copies of the DVD and shared them around, asking only that people watch, then pass the DVD on to someone else or return it for me to give out again. People watched it (or so they told me), a few even shared it. I have no idea whether it changed any minds or stimulated any actions, but at least I had done *something*.

More and more books came out about climate change, including Elizabeth Kolbert's excellent 2016 Pulitzer Prize–winning book *The Sixth Extinction*, conveying the science, the statistics, the global consequences of climate change.

But what I knew from a long career writing about wildlife and conservation is that we all relate to things best through story. Data and science reach the mind, but story reaches the heart. Story shows how science affects the lives of regular people. A story, if it's a good one, grabs our attention and makes us listen.

What I have to share with readers, then, is my story. My climate change story. It's the story of what my family and I see happening at our special place, our piece of the wild in the foothills of the southern Rocky Mountains. This book is not a primer to climate science—I'm not a climatologist and there are many excellent, detailed, and authoritative sources on climate change—and it doesn't offer magical solutions.

In the way of all great stories, ours stands in for the stories taking place in one way or another in the lives of everyone. The things happening on our land and touching us are happening also in the lives and backyards and special places of real people everywhere. And what we all want is for our climate change stories to have happy endings.

1

BLUEBIRDS
IN THE MEADOW

1995 is the hottest year on record, since global temperature
records started being kept in 1856.
—British Meteorological Office
and the University of East Anglia

The balance of evidence suggests a discernible human
influence on global climate.
—1995 *Intergovernmental Panel on Climate Change*
Second Assessment Report

August 1995. We move through the meadow, dodging patches of
prickly pear cactus. The seedheads of blue grama grass, curved like
eyelashes, brush my jeans with a gentle *whish-whish*. Grasshoppers on
spring-loaded legs erupt from the meadow at odd angles ahead of us,
spreading their pale-edged wings like dark cloaks. With our approach,
lesser goldfinches flutter from the seedheads of wooly mullein, whose
spear-like stalks stand in ranks across the meadow as if left by some
ancient warrior legion. The spicy scent of sagebrush follows us like a
faithful dog.

My husband, Rick, kicks at a pile of elk scat, then turns and smiles
at me. *Elk poop*, I know he is thinking. *Could be the dealmaker.* He has

1

made the point more than once that I am the only person he knows who gets excited over animal droppings.

And I am excited. Where there is scat, there are elk. Their antlered shapes likely fill this meadow in fall and winter, the half-ton bulls sparring antler to antler, competing for females who drift between attractive bulls to choose the bull they will mate with, rather than the other way around. Autumn evenings here would be punctuated by the shrill bugling of the bulls—eerie, hollow shrieks that seem oddly high-pitched coming from such large beasts.

I had convinced Rick to come on this late-summer Sunday outing to look at land with me. Married the previous March, we were drawn to each other by our mutual love of wild places and wild things. I often describe Rick as Mr. Outdoors. He had run the outdoor program at the YMCA's Snow Mountain Ranch, near Granby, Colorado, hiking, climbing, and exploring throughout the mountains. He has reached the summits of all fifty-four Colorado "fourteeners," mountains rising to fourteen thousand feet or more in elevation. I am a zoologist and professional nature writer who has hiked, camped, backpacked, birded, climbed, and explored Colorado my whole life. My entire being is committed to the natural communities of my state and the American West.

As a kid I'd never had a permanent home, but Colorado had been my anchor. I had grown up as an army brat, moving ten times by the age of seventeen. Between all our relocations and school changes, we'd come every summer to my grandparents' log cabin in Estes Park, Colorado, on the eastern border of Rocky Mountain National Park. My childhood summers spent running wild on the mountain, with hummingbirds whistling around my head and mountain bluebirds flitting across the meadow like scraps of the sky, had determined the course of my adult life.

Through all those childhood moves, Colorado was my polestar, the place I could come to every summer, the place I felt most at home. *I will make my life here*, I decided as a kid, and I did. I enrolled at Colorado State University, earned a degree in zoology, and made a career as a wildlife and nature writer. Though the family cabin in Estes Park was long gone, I was determined to buy land as an adult and create a similar refuge to be close to the wild.

Rick also had a special childhood place that was a part of him. He'd spent his boyhood summers at his grandparents' cottage on one of the Finger Lakes in upstate New York, swimming in the icy waters of Skaneateles Lake and exploring the steep hillsides above the water.

So it wasn't hard to convince him we should create a place like that for us, a weekend retreat far from the city where the landscape and wild things would renew our spirits. Where a child could run wild, get dirty, scrape knees, climb mountains, chase chipmunks.

Land in the central Rocky Mountains west of Denver—close to ski resorts and glamour gulches for the rich and famous—was much too expensive for regular folks like us. Then I saw an ad in the Sunday paper for land in southern Colorado. The price per acre was reasonable. So south we drove, to western Las Animas County.

We've spent the day exploring land for sale and have narrowed our choices to two parcels. Rick likes this thirty-seven-acre piece of land among the rugged foothills west of Trinidad, only about seven miles north of the New Mexico state line. Broken by rock-tumbled canyons and streaked by arroyos that lie bone-dry most of the year but rage like angry bobcats when rain falls, it is a land of grassy meadows ringed by blue-green junipers and piñon pines. The landscape feels more like the American Southwest than it does the Rocky Mountains.

This plant community known as piñon-juniper, or PJ, is nicknamed the "pygmy forest" because the trees rarely grow taller than forty feet. Low stature notwithstanding, these trees will bear rich crops of piñon nuts and juniper berries and support legions of insects, all providing rich food for the birds and animals that mean so much to me, which are the basis of my career as a nature writer. In areas where there is more water, ponderosa pines tower above the landscape.

But I wanted land in the high mountains—cloud-brushing, ethereal acreage of high peaks that might have snow well into summer. This morning I fell in love with a thirty-five-acre lot on top of a mountain, with a view to the west of the high peaks of the Sangre de Cristo Range—the "blood of Christ" mountains. And to the northeast, rising from the prairie in the morning sun, a view of the twin mounds of the Spanish Peaks, known to the Utes as Wahatoya—the "breasts of the earth." I pictured myself writing away on that mountaintop, its slopes tinged blue by a forest of white firs with needles as soft and round-tipped as the wing feathers of an owl.

Now I stand in the meadow of the lower elevation land and turn a slow 360 degrees. The view is to the east, where the sun will be born each morning behind the long, flat expanse of Raton Mesa. Clouds skiff above the mesa in an impossibly blue sky, and I have a sense that I could surrender myself to their elegant promenade and there would be no passage of time, only these castles of water vapor carrying me with them.

From the creek that carves Long Canyon a half mile to the east, the land sweeps gently up Toro Canyon to the summit of Montenegro— Black Mountain—rising to seventy-two hundred feet behind us. The tubby shapes of Colorado piñon pines, as wide as they are tall, ring the meadow, mixed with the drooping, silvery branches of Rocky Mountain juniper and the stiff, yellow-green of one-seed juniper. Peering over them, like NBA players standing among ordinary mortals, are stately ponderosa pines.

The flight of birds across the meadow draws my attention, as it always does. A flash of blue in the sunshine—a cloud of bluebirds, I realize, sparkling across the meadow. Not sky-blue mountain bluebirds, but birds that gleam purple-blue in the sun.

"Western bluebirds," I say to Rick, not taking my eyes from the birds. I raise my binoculars, track them as they alight like blue ornaments on the shrubby piñons and mullein stalks. One bird flies suddenly upward, then stops in midair. Helicoptering above the grass, tail dropped, wings rowing frantically forward and back, forward and back, it hangs poised in one spot for impossible seconds.

Then like a stooping falcon, the bluebird drops to the grass, stabs at something with its bill. Then it is airborne again, the sage-green legs of a grasshopper projecting stiffly from its beak. I wonder vaguely if it is one of the 'hoppers we had flushed, sent by our passage into the beak of death.

The bluebird lands on the branch of a nearby piñon next to an awkward-looking bluish bird with a speckled breast. Like their thrush-family cousin, the robin, the young of bluebirds have speckled breasts. This is a newly fledged bluebird from a clutch laid late in the season, out of the nest but not ready for prime time. The baby is the size of the adult, but its fluffy shape and take-care-of-me posture are undeniably those of a young bird. It doesn't yet understand that with its flight from the nest, the world has taken an immense turn and mom and dad will not bring home the groceries much longer.

The baby opens wide its yellow bill and the adult pokes the grasshopper into the gape. Gulp. Immediately, the yellow bill flares wide again—*Feed me, papa!* Off the adult goes, back to the store, then back home with an insect. Gulp. The bill yawns wide again, the yellow-rimmed gape a visual demand the adult bird is compelled by instinct to fulfill.

After the fourth feeding, the adult flies off and does not return. The baby sits, waiting, trying perhaps to grasp this turn of events—the great world spread before it, plenty of food hopping around the grass but no adult to catch it and deliver. The adult's continued absence seems a clear message: *Get a job, son!* I almost imagine a tear forming in the corner of the fledgling's eye. We all know it's hell to grow up.

Finally the baby launches and flies off after the parent. I look at Rick. He is smiling. He knows he has me, that I will let dreams of the mountaintop go and choose this place. The bluebirds have gone to bat for his team, and they've hit one out of the park.

CABIN JOURNAL—OCTOBER 1995: *An enormous group of pinyon jays blows in from the southwest, a boisterous blue posse. Easily one hundred birds, maybe three hundred or more.*

The pinyon jays tumble across our big meadow in a milling crowd, landing on the ground and in the pines, foraging, calling to each other. Some hop around and check out our new fire ring. There is nothing shy or tentative about this crew. They are like a group of Hell's Angels, shouting, gunning motorcycle engines, descending on a small-town saloon in loud and raucous disorder. Heedless of the other patrons, they usurp all the stools, pound on the bar. The smaller birds, wary of these rowdy intruders, flee for quieter perches. But the rowdies don't stay long. After pausing in the piñons to pry what nuts remain from the cones, they head out again in a straggling flock. They're back in the air, on their bikes, heading out of town in a confused, milling riot to more fertile pickings. We hear them calling back and forth in their three-tone, nasal mewing that drops in pitch—*mew-ew-ew, mew-ew-ew.* They surge in a loose confederation across the lower meadow and over the ponderosas to the far ridge, then beyond, their mewing fading, fading, until they are gone.

I've gone all Thoreau, hunched over a small notebook with a rigid dark-green cover, brushing away the pine needles the breeze drops on my page. It's not a proper nature journal—no tooled leather cover, heavy vellum paper, or brass clasp. Just something I grabbed at the last minute as we headed out to camp on our new land. But it will do.

This is our third camping trip since we purchased in August. We set up our little backpack tent in a flat, grassy spot sheltered by ponderosas with a view of our small pond and Raton Mesa beyond. We've brought the luxury of lawn chairs and a camp stove-top espresso maker. I doubt Henry David had either of those when he lived at Walden Pond. But I'm entitled to a few indulgences.

Our first few visits, I'd scribbled notes and sightings on scrap paper, recording our observations more as an afterthought, really, outshone by the first excitements of discovery as we explore this new place. These first records are Spartan lists of species:

Birds
Western bluebird
Mountain bluebird
Pygmy nuthatch
Black-capped chickadee
Mountain chickadee
Common raven
Scrub jay
Plain titmouse
Sandhill crane (in flight)

Pine siskin
Lesser goldfinch
Northern mockingbird
Mallard
Common nighthawk
Wild turkey

Mammals
Mule deer—sightings, scat, bones, antlers
Elk—sightings, scat
Black bear—lots of scat, bear rub tree with coarse black hairs in sap
Mountain lion—tracks in mud of runoff seasonal stream after rain
Coyotes—scat, lots of evening and night howling
Rock squirrel—in piles of rocks
Pocket gophers—mounds
Bats—feeding in evening over pond meadow; droppings (guano)
     among rocky cliff (big brown bat?)
Black-tailed jackrabbit
Cottontail (Mountain?)
Colorado chipmunk

Herptiles/Invertebrates
Chorus frog
Eastern fence lizard
Short-horned lizard
Prairie rattlesnake
Wolf spider

Keeping a nature journal is a hallowed tradition among nature writers. We record what we see, with identification of species and details of size, shape, behavior, season, weather, moon phase, time of day. But I want do more than record observations. "I went to the woods because I wished to live deliberately," Henry David Thoreau wrote in *Walden;*

or, Life in the Woods, "and see if I could not learn what it had to teach, and not, when I came to die, discover that I had not lived."

I, too, want to live closer to the land and learn its lessons, to get to know our piece of the wild in all its incarnations. Already the land is retracting into the sleeping season as wildflowers fade and plants pull their vitality back into their trunks and roots. Wildlife is beginning to migrate, hibernate, or survive however they are adapted to do. I will journey as they do through the progression of the seasons, as wild things awaken in spring ripe with life, bring forth seeds and nuts, hatchlings, and cubs. Watch the earth inhaling and exhaling over and over through the seasons, the years, the decades.

A flutter of activity in the meadow just in front of me draws my attention. With departure of the pinyon jays, the locals are emerging, and a group of small songbirds wings across to settle on a fat piñon to my left. Five male western bluebirds, their plumage glowing purple-blue in the light, with breasts of brick-red, bob on the branch tips like Christmas tree ornaments.

I bend my head to my journal and continue writing.

2

NEST BOXES

Almost all scientists now consider the evidence for global warming to be "incontrovertible," and many believe it is accelerating at an alarming rate.... This year is set to be the hottest of the millennium.
—*The Guardian*, November 14, 1999

The summer night holds us in a soft embrace, the air caressing our faces, filling our lungs with cool, clean breath. We're lying in the long grass of the big meadow upslope from our campsite, peering through the blue grama like resting lions. We've walked up to enjoy the magic of the evening. This meadow lies north to south across the slope, sweeping gently downhill toward Long Canyon. Some ten miles distant, across a high and wide space of air, the profile of Raton Mesa rises in silhouette against the eastern sky. At our backs, Elk Ridge rises to a saddle, meeting a series of wooded ridges that climb ever steeply to the top of Montenegro. Above the peak, the Big Dipper hovers, its bowl pointing toward the North Star, faint but constant over the ridge to the north, the one we call Rattlesnake Ridge for the many times we've encountered rattlers up there. Spinning around this pole star in slow but infinite rotation, the Little Dipper stands at this moment on its handle, emptying itself to the Earth below as it does every night in its endless circular journey.

The dogs range around us, sniffing out entire worlds we cannot detect. Their wanderings stir the spicy scent of fringed sagebrush, which comes to me across the blue air, a scent even my poor nose can recognize. We call the dogs to us, settle them next to us. Lit by the glow of a waxing gibbous moon, so bright it casts our shadows across the grass, we are comfortable here. Like night creatures, we breath in time with the evening, spectators but also part of the scene.

Around us, the night world is a wondrous living thing, a humming symphony—the trill of crickets, the clunky drone of a passing beetle, the far-off hoots of a great horned owl, the *peent, peent* calls of high-flying nighthawks. Bats flicker above us, fluttering, diving, pirouetting in midair like ballerinas. They perform in silence, but I know that if we could hear in their ultrasonic range, the night would be filled with their bellowing clicks and pulses as they cast sonic nets in search of flying insects, reading the echoes of their calls bouncing off their prey.

A dark shape wings suddenly past our heads. *What the heck was that?!* The phantom moves away, flying just above the grass. Then a *poo-wee-ee* call, and we recognize a hunting poorwill, cousin to the nighthawk, but a night hunter who stays much closer to the earth. It shelters in wait on the ground, flying suddenly up to grab low-flying insects, or hunts across open land just ten feet above the ground. Its shadowy shape soon disappears, but we track its flight around the meadow by sound—a call to our left, another call closer, a pause, then a third call from our right. Poorwills are familiar neighbors. We've heard them many times at night from spring through fall. One night as we drove in, the truck's headlights sweeping the driveway, I saw two glowing red lights on the ground at the edge of the road—the night shine of a poorwill's eyes reflecting the light. With beaks that gape wide, poorwills gobble up quick-moving insects, their mottled plumage so wonderfully camouflaged they disappear like avian magicians when they land on soil or gravel.

It's as if the poorwill has carried a revelation. "This is where we should build," I say suddenly. Up here in this wide meadow, with its clean sweep down to Long Creek then up to Raton Mesa, feels balanced and inviting. We've seen so much wildlife in this meadow the last three years—heavy-antlered elk, mule deer with elegant heads and slender legs, plume-tailed coyotes drifting through on silent paws. Birds of all size and shape and color. Even tiny chorus frogs hatched in ephemeral potholes filled by spring rains. A cabin at this spot would have a wonderful view and an open vista of sky. We've spent months and years walking our land considering where to build—near the pond at our longtime campsite, along one side of the meadow we've started calling Elk Meadow where it spreads upslope, long and narrow like a ski run. But nothing ever felt quite right. Now the best site is revealed to us by a common poorwill.

Rick agrees. Centered in a wide meadow across the slope, with a gradual downhill grade, it feels balanced and pleasing, the perfect spot. I feel like Brigham Young arriving with the first Mormon pioneers in the Salt Lake Valley and pronouncing, "This is the place."

And now, like Brigham and the Saints, we must build in the wilderness.

The Coleman lantern casts a warm light across the log home catalog I am studying. I peer closer to make out the details of a floor plan for a simple rectangular cabin with two bedrooms, one bath, an open living-dining-kitchen area and a covered porch. It's chilly on this September evening, and I've thrown a blanket around my shoulders as I sit at the picnic table in front of the tent. Across from me, Rick studies another catalog. Our baby, Olivia, born in early 1998, nestles asleep in layers of blankets on his lap.

The evening's flying creatures must be interested in log cabins too, because moths and flying insects of all sizes and shapes flutter and dance with the lantern light. A low drone like the engine of a small plane moves slowly and deliberately past my head and straight for the lantern. A black beetle at least two inches long, flying like a tin can sputtering on empty, collides with the lantern glass and tumbles to the table on top of the open catalog. As I watch, the beetle struggles to right itself, six legs pistoning. Finally it tips slowly upright, closes its wing covers with a click, then clambers around the slick page as if reviewing the floor plan. It stops over the space for the living area, not seeming in any hurry. Is it imagining how it might arrange the furniture?

Finally the beetle opens its wing covers like an umbrella, the second pair of wings sputters into action, and the beetle lifts off, droning slowly away into the darkness like a bumbling professor in a homemade plane of spare parts.

I look over at Rick and smile. "That settles it," I say. We will go with this floor plan. The beetle has spoken.

For the nearly four years we've camped here on weekends and school breaks, we've been content with our rustic accommodations. From our small backpack tent, we quickly moved up to a roomy dome tent seven feet tall in the middle and big enough for a rug over the nylon floor (luxury!), with an awning-covered "front porch." But over the years our camp grew more elaborate—a "pup tent" for the dogs, a dining canopy over the picnic table, a fold-out camp kitchen (thank you Coleman) beneath a tarp, a hammock strung between the trees. Idyllic, except it all needed to be set up and taken down each visit, wearing us out before we could relax. The arrival of our daughter, Olivia, in 1998, which led to nursing a baby in a tent on twenty-degree mornings in spring and fall, was the tipping point. It was time to build.

Dreaming about what we would build had occupied many pleasant days since purchasing the land. Early on we accepted that living full time here was not realistic and we just wanted a comfortable weekend and vacation retreat. We considered everything—a prefab cracker box little more than a glorified Tuff Shed, a canvas yurt, a straw-bale home, a log-sided modular that looked like a suburban home on the inside, even a full-blown second home that would be larger than our house in Denver. Finally we settled on a log cabin. We began visiting log home shows, subscribing to log home magazines, sketching floor plans. We debated handcrafted versus milled logs, full round profile versus D-shaped, saddle notch corners versus butt-and-pass.

At first we dreamed big, then crunched our finances and dreamed a bit smaller. Finally we settled on a 768-square-foot two-bedroom log cabin package from a company in New Hampshire. We tried to find a Colorado, or at least western, company to work with, but the costs, plans, and inclusions were never right. Our package included everything to "dry in" the cabin, from the subfloor to the green metal roof, including doors and windows and knotty-pine paneling for the interior walls and ceilings.

To afford anything, though, we would have to do much of the building ourselves, with the help of every friend and family member we could convince to sign up. And we were up against a tight time frame. Rick, a high school teacher, taught on a year-round calendar, limiting our building season options. He would be off for three weeks in June—the prime month to build. Before we could set the first log, though, there was the massive job of planning and steps just to get ready—driveway, foundation, electricity, cistern and septic install, log delivery—and rounding up as many willing friends as possible to help in our old-fashioned cabin raising. Could we do it?

At our house in Denver, the golden leaves of the backyard cottonwoods were letting loose their moorings and drifting to the ground. Winter birds like juncos and robins crowded the feeders and birdbath. Earth had moved past the fall equinox and was heading for winter solstice, and our window to get all the planning done was closing along with the season.

On the kitchen table in front of us, log home plans, contractor estimates, and sheets of paper scrawled with numbers and tallies and dollar figures spread across the wood. Rick was looking at the calendar for the upcoming year, 1999, with the dates of his school breaks circled. He shook his head and set down the calendar. "Let's build a year from this June."

Maybe it was the spiked cider we were drinking on a chilly autumn evening, but a "go for it" sentiment surged through me. I looked at him. "Why not build THIS June?" We stared at each other, both of us wide-eyed with a *Do we dare?* expression, our minds racing back and forth between eagerness to get started and concern that there was no way to get the planning done in time. Then Rick, bless his usually cautious heart, tossed the calendar down on the table and grinned. "OK, this June."

Spring has summoned a burst of activity on our land. It's the awakening season, and the birds and wildlife are definitely awake. A male western bluebird wings past carrying strands of grass in its bill. A female flies by with more nesting material. Mountain bluebirds are busy too, disappearing into the edge of the piñon pines where they are building a nest in some mysterious tree cavity. Pairs of red-tailed hawks dance in the air. Ravens crisscross the sky, long sticks trailing from their bills. We record mountain chickadees, ash-throated flycatchers, red- and white-breasted nuthatches, white-crowned sparrows, northern flickers, hairy woodpeckers, solitary vireos, and scrub, Steller's, and pinyon jays. A chipping sparrow trills its staccato song from atop a fat piñon. A pair of robins gathers grass for a nest they are building somewhere up the slope.

Everyone is building nests and homes. Now we prepare to do the same.

The first step is a big one. The driveway is cut and graveled from the road up Elk Meadow to our building site. But the fresh scar on the land grabs at my heart. It is like a wound, a slash in the skin of the Earth. And it is just the first of the alterations we will make. Of course, we cannot come to this place and build without leaving a heavy mark. I glimpse what our footprint here will be, and I must own it. As we make this place more habitable for us, I grieve the loss of its wild nature, and I turn, as humans always have, to ritual. As Rick gets busy with tasks, I gather a small votive candle and carry Olivia with me to a special spot at the edge of the meadow. It is enclosed by piñons but open to the east, a place to view the sunrise. I hand Olivia a piece of sandstone, which she happily waves around as I stack a small windbreak of stones on a flat rock set where the land drops steeply downhill. I set the votive in the lee of the windbreak, arrange a few piñon cones, a bit of bone, and a twisted sprig of fringed sagebrush. Eager to join in, Olivia adds her stone as I tell her how this is our way of thanking the land and the wild things for sharing this place with us. Her wide hazel eyes gleam with interest and excitement. She will follow her own path in life, perhaps far from these rough mountains to bright lights and a big city, but I hope she will carry with her a respect for the natural world, the knowledge that what happens to it also happens eventually to us.

I light the candle, then gather my baby onto my lap. I thank the divine for the gift of this place, thank the landscape and all the living things, pledging to be a good steward of this land and nurture it as I can. Carefully I hold the plume of sagebrush in the flame until it smokes. I set it on the stone and draw the fragrant smoke to me, swirl it around Olivia as she laughs and waves her pudgy hand through it too. A gentle breeze carries the purifying smoke up and beyond us, out across the land.

With our driveway graded to the site, a foundation poured, and electricity, cistern, and septic in place, we are ready to build. For the next three weeks our lives will be part *Walden*, part *A Year in Provence*—living simply and close to nature and with a community of friends coming together in the spirit of fellowship to help us build our cabin. And like the couple rebuilding a villa in Provence, we are under a tight deadline to complete the task!

We've set up a tent city at our campsite with the large tent, a guest tent with sleeping bags, a "kiddie corral" on a grassy patch, an expanded camp kitchen, a picnic table under a canopy, lawn chairs around our rock firepit with a view of Raton Mesa, and space for more tents and tent trailers. The cistern is full of water, a porta potty is in place, a camp shower hangs where the southwestern sun will

heat the water to plenty-warm by afternoon, and a power line now
marches up the driveway to the building site. (We had investigated
solar but the cost was way outside our budget.) It's early June, and
we have three weeks to stack logs into a cabin, put on the roof, and
install windows and doors so the structure is "dried in" and secure. So
many things have to fall into place to meet that schedule. How many
of the friends who committed to help us build—"Build a log cabin?
Like in *Little House on the Prairie*?! I'm in!"—would really show up
when we needed them?

The morning arrives and we are up early. Like Ma on the ranch,
I make eggs and pancakes on the two-burner Coleman for our first
crew of friends who have come to help. Two neighbors, John and
Mel, arrive in cowboy hats and well-scuffed boots, ready to work. We
have hired them for tasks beyond our ability. In the grand tradition of
fresh starts and new identities in the wide-open West, both are white-
collar professionals from other places and with formerly urban lives who
reinvented themselves here in this rural county. John, a self-described
"bean counter" from Chicago, now has a contractor's license and will be
something of an overseer. Mel, a computer scientist and college professor from Iowa, is a sort of odd-jobs guy with a barn full of equipment
and tools, bringing his geek brain and meticulous attention to detail to
even the simplest task.

The grinding of a semitruck sounds in the distance, ebbing and
flowing as it travels up a rise, down an arroyo, drawing nearer. Finally,
our cabin arrives. Or rather, the components arrive, neatly bundled on
the back of a single flatbed trailer like a life-size set of Lincoln Logs.

I can practically hear the music from the Amish barn-raising scene in
the movie *Witness* as we get ready to set the first course of logs into
place. The Colorado sky is an endless ceiling of blue, and we squint
from the southwestern sun, which is comfortably warm but not yet
blazing. Early summer rains have drawn the meadow grass thick from

the ground, interspersed with banners of purple and white penstemon, red fairy trumpets and Indian paintbrush, clumps of white peavine. Silvery plumes of sagebrush release their spicy scent when any of us brushes past. Birds fly purposefully above, around, and past, busy building their homes and feeding their families, though we are too busy building our own home to take much notice. Our voices call back and forth—"Ready with the first log?" "Grab that handful of spikes," "Can you bring me that small sledgehammer?"—joining the occasional calls of birds and insects, the breeze in the pines, all the wonderful music of a June morning in Colorado.

Our Chesapeake Bay retriever, Jasper, sleeps in the shade of one of the log stacks as Buddha, our friend Paul's young Labrador, tries to pester him into play. Soon we will eject the dogs so we can turn those logs into a house. Down at the campsite, my sister, Sally, is on kid duty, loving the chance to play with her niece, Olivia. She'll enjoy the logs once they're a cabin, she says.

We will stack the logs in "courses," basically layers, on the newly built subfloor, starting at the southwestern corner, then working in a counterclockwise direction around until we can set the last log in the course butted against the first. The logs have come numbered and lettered—the numbers reflecting the sequence each log will be laid, the letters denoting the course. The four walls of the cabin will rise in courses A through N. The gable ends will be set on the south and north walls, tapering up to the single, triangular log—only one foot, four inches long at its base—we will set carefully at the apex like the keystone in an arch.

We opted for milled D-profile logs, which are eight inches high, six inches wide, flat on the interior side and rounded on the exterior (hence the name). Each has a raised "tongue" on its top side and a long groove in the bottom that will fit over the tongue of the log below. We've drilled holes vertically every thirty inches along each log. We will hammer metal spikes through to secure each log to the one below.

Rick and Paul each grab an end of log A1, the first log of the first course, which is nearly fourteen feet long. As they lift it, a male western bluebird flies across the meadow, a tuft of grass in his bill. Our building materials are heavier and more unwieldy than the grasses and old feathers I see the bluebirds carrying for their nests. On the other hand, they have to rebuild every year. Our logs are a lot more durable.

Rick and Paul wrestle the log into place, and I get the honor of driving the first spike, basically a twelve-inch nail I whale on with a

small sledgehammer. *Toing, toing, toing,* the metal sings out as I bash away until the spike is buried flush in the raised tongue along the top of the log. There is much cheering and hooting from the crowd, as if we were driving the golden spike connecting the first transcontinental railroad instead of setting one log on a little cabin. I straighten up, grin, and pump the hammer in the air with two hands like Thor as everybody cheers. Then I transform into the crew boss. "Get to work, you bums!" I yell, and we're off to the cabin-raising races.

———

CABIN JOURNAL—JUNE 1999: *Found a mariposa lily by the house site, three creamy petals forming a yellow-centered cup that holds droplets of rain. Not much time to look at birds, but serendipitous sightings include western and mountain bluebirds, mountain and black-capped chickadees, plain titmice. Western kingbirds everywhere along the roads. Ravens dance above Rattlesnake Ridge, and turkey vultures spiral upward on rising air currents. Ash-throated flycatchers very active. Robins, solitary vireos, scrub, Steller's and pinyon jays, cowbirds, Brewer's blackbirds. We hear red-winged blackbirds, a western screech-owl, lots of chipping and white-crowned sparrows, many broad-tailed hummingbirds at feeders and displaying. Lots of nighthawk and poorwill activity. Several Woodhouse's toads at our house site. I love these ungainly, bug-eyed toads, warts and all. We hear chorus frogs at many puddles in the meadows and the arroyo above the campsite and around the cabin. I spot tadpoles in the shallows of the pond. Bats show up every night, hunting the air above our campsite in their fluttering, erratic flight. They seem particularly busy over the campfire, and I wonder if the light and heat attract moths and night-flying insects. Does the smoke confuse the insects, maybe slightly anesthetize them so they are easier prey for a bat?*

———

By the end of the first week, the log walls form a roofless cabin, with open spaces for windows and doors. Now we will move on to framing the interior walls and putting up the roof rafters, decking, and metal roof. Rick and I stand in what will be the living room and look up at the sky. It's impossibly blue, inviting me to rise up and float in its vast space like some infinite being. The log walls already set up a barrier between us and nature; I can't stand to close us off from that stunning western sky. I loop my arm through Rick's. "Let's just not put on a roof!"

He laughs. "There's a plan. And think how much time that would save us. No problem making our deadline. But there is the problem of that 'drying in' thing."

"Oh yeah, that," I say.

But before we get the roof on, nature sends a message reminding us of that "drying in" thing.

It starts raining. This isn't the gentle, warm rain Gene Kelly cavorted through in *Singin' in the Rain*. Colorado rain is cold even in midsummer. The air temp drops to the low fifties, and we bundle up in sweaters, knit hats, and our Gore-Tex raingear. Water drips from the pines, sags the awnings on our tent and the canopies over the picnic table and camp kitchen. It streams down the trough along the driveway before diverting into the pond, where the water level rises. For two days we're not able to work on the cabin. Our helpers leave, and I call friends who would be heading down next to put them on hold until the weather clears.

We carry Olivia and walk up to the cabin, blowing plumes of steam with each breath. We've laid planks for walkways from the gravel drive-way to the cabin, like a frontier western town before paved streets. Water pools on the subfloor and puddles into mud all around the cabin. Now I wish the roof, the one I said maybe we didn't need, were already in place.

If our building deadline weren't haunting us like a specter, we'd welcome a little downtime. We're tired from long days of hard work. Our hands ache from gripping and swinging a heavy sledgehammer. Carrying, drilling, pounding. Rick complains that he hammers spikes in his sleep. A few days' rest would be great, but there's no time for that. We were already just keeping on deadline; even a day's slowdown could throw us off. Rick has to get back to teach in a week and a half. What will we do if we can't catch up? Will he head home to Denver while Olivia and I stay here and try to get things finished? Could Ma Ingalls have finished the little house on the prairie by herself, with a baby on her hip?

Back at camp we cook a warm dinner, then cuddle in the tent with Olivia to read stories and hope for a change in the weather.

Like flipping a switch, the rain ends in the night, the next day dawning with a sky of stunning blue and a sun so inviting it is as if the Earth were smiling. Rain droplets glitter jewel-bright on the grass, and a mist rises ghostly from the meadow.

At the edge of the pond, newly swollen by rain, a great blue heron poses, motionless, intent on hunting our chorus frogs. As I watch, it lashes out with coiled neck and stabbing bill, striking lightning fast into the water.

I didn't see the heron arrive this morning. It was just there when I woke up, materialized to take advantage of the bright new day. Life moves forward and we all respond. The next moment I am on the phone to friends—*Come on down, we're back in business.*

Midsummer's Eve. Five of us sit in lawn chairs around the rock firepit, beers in hand, staring into the flames. The days are long at the summer solstice, and we take full advantage of the daylight, often not calling an end to the workday until 8:00 PM. We're all a little glaze-eyed and tired after hours of physical work. But it's a good tired. We spend our days in the crystal air and golden sun, working with our hands and our bodies, getting covered with sawdust and dirt and sweat. A good sweat.

Our cabin-raising crew has constantly cycled, friends arriving to help, working a few days, heading home as others show up, the ebb and flow always bringing enough hands for the task. After ten-hour workdays, we bring well-earned appetites for our shared meals of spaghetti or anything that can be cooked for a group on the double-burner Coleman. Tonight we crowded the picnic table to share a pot of stew, bowls of salad, and a mound of bread, an echo of the American tradition of feeding hungry farmhands at harvest.

Comfortably full and ready to rest our muscles, we enjoy an evening of communion and goodwill around the campfire, more magical than any Midsummer's Eve imagined by Shakespeare.

A rasping scream sounds from the darkness, distantly, from up the arroyo where the mountain steepens and cliffs rise from folds of broken rock. We look around at each other, faces glowing in the firelight. "Was that what I think it was?" someone says. Mountain lions are regulars in this neighborhood—we've found tracks and scat over the years—so I'm not really worried about hearing one tonight. They're always around, just unseen. But I wonder how many of us will forego a midnight trip to the porta potty.

"In the jungle, the mighty jungle," Rick begins, "the lion sleeps tonight." We all join in, "Aweem-away, aweem-away . . . " The sound of our camaraderie twines with the smoke and the flames, rising among the pines toward the glittering night sky. The eagerness of our friends to help us build, their joy at pitching in and working so hard, amazes me. They are excited to be part of making this cabin rise, enjoying the feeling of accomplishment and satisfaction after every exhausting, sweaty day. We are gratified beyond words for the support of our great friends. This is human community, the way a vulnerable species—weak, slow, without claws, sharp teeth, or protective fur— has come to dominate the planet. We don't have many natural tools, except for a big brain and each other. *Homo sapiens* can accomplish amazing things when we support each other and work together. When those accomplishments have unforeseen and calamitous consequences, the power of working together also becomes our greatest strength for meeting future challenges.

Earth turns inexorably through the summer days, carrying us toward our deadline. The finished roof now gleams a metallic green against the pine forest and the interior walls are framed in, but there is still so much to do! We hustle to install the doors and windows, then the trim around them so they are weather tight.

The last day arrives. Our friends and helpers have headed home, and soon we will too. Rick is up at the cabin finishing the last critical things before we can leave, cleaning the site, buttoning up the place, making sure it is safe from rain and animals and any intruders. His last task is to install the door locks.

I've begun the huge job of closing up camp, dropping the tent, dismantling the kitchen and canopies, cleaning up anything that might attract bears or leave the site messy. I pack up everything to either take back to Denver or store in the cabin, where we will now "camp" inside while we spend weekends and school breaks through the end of the year finishing the interior on our own and with the occasional help of friends.

The bluebirds we saw carrying nesting material a few weeks ago have completed their nest-building too. When I see them now, they are hunting grasshoppers to feed a hungry mate or newly hatched babies.

Finally the cars are packed, our two vehicles loaded to the brim with more tied on top, looking a little like the Clampetts heading for Beverly Hills. Holding Olivia's hand—she is almost, but not quite, ready to walk—we stand inside the cabin for one final look. The floor is still only rough plywood. Raw two-by-four framing delineates what will eventually be rooms, though right now we can see right through the entire place from one exterior wall to the opposite. We've done it. We've built our cabin, have it dried-in and secure, so we can leave.

But before we lock the door, I have one last task. I bring four empty wood boxes inside to leave behind, each with a one-and-a-half-inch entrance hole in one side. Bluebird nest boxes. We have built a wood nest box for ourselves. By the end of the year, I will put up these boxes around our meadows, ready for the bluebirds to use when they return next spring.

"OK, let's hit it," Rick says. We lock the doors for the very first time and drive away, the cabin disappearing behind us among the trees.

———————

By the summer of our cabin building, we had been keeping our nature journal for three and a half years, noting wondrous things. Courting red-tailed hawks tumbling together through the air, joined talon to talon. Wolf spiders carrying dozens of tiny babies on their backs, hunting from burrows secreted in the grass, with one-inch-wide entrance holes so perfectly round they look drilled by a machine. The list of bird species we spot on our land and down along Long Creek keeps ticking up and up, reaching ninety-six species by spring of 1999.

By the late 1990s, I had read of concerns about global warming, but in my professional life as a nature and wildlife writer, I am still focused on the conservation of specific wildlife species, the loss of habitat, the development of wild places. The voices warning of climate change are still faint.

Far from noticing any climate-related changes on our land, we are still caught up in the wonders of discovery. The natural community here seems incredibly rich and vibrant, the meadows bright with wildflowers, birds, and other wildlife moving across like players on a stage. Each new visit brings anticipation for what birds we will see, whether we'll hear coyotes or find bear tracks. It will be years before we have enough of a record to recognize changes beyond the cyclical ones that are part of the system. When we do begin to see puzzling new patterns or species, we won't quickly recognize the causes. But while we remain happily unaware, the changes of a warming climate are underway.

3

PINKIES

Plant and animal ranges shifted poleward and up in elevation
for plants, insects, birds, and fish. . . . Earlier plant flowering, earlier bird arrival, earlier dates of breeding season, and
earlier emergence of insects in the Northern Hemisphere. . . .
Growing season lengthened by about 1 to 4 days per decade
during the last 40 years in the Northern Hemisphere, especially at higher latitudes.
—*Climate Change 2001 Synthesis Report,*
    International Panel on Climate Change
    Third Assessment Report

A scrap of blue flits across the meadow, swooping to a perch atop
a dry stalk of prairie sunflower. I smile. The bluebirds are back.

From his bobbing song perch, the male western bluebird's fluting
music dances across the grass, singing of joy and sunshine and awakening. Who wouldn't sing with joy on a spring morning like this one?
He's back from wherever he's spent the winter—most likely the lower
elevation valleys and canyons of southeastern Colorado carved by the
Arkansas River on its journey out of the state—and he's looking for love.

The sun burnishes the male's plumage to cobalt, his breast brick
red. I know it's a trick of the light, that deep blue color. It doesn't
come from pigment but from light reflecting off the intricate structure
of the feathers. Perched in the shade, this same handsome fellow will

appear a drab gray. But if his stunning color is nature's subterfuge, I'm happy to be tricked!

In a moment, a female bluebird, her plumage a drab gray tinged with blue, perches on a nearby sunflower. Is she drawn by his song, come to check him out, or have they already bonded, either earlier this spring or even last year? Did they nest here the previous summer, when we were busy building the cabin, flitting past us as we hefted logs and hammered spikes?

As I watch from the cabin's porch, the pair flies into the branches of a piñon at the edge of the meadow. Attached to the trunk is one of the bluebird nest boxes we put up over the winter. It faces the grassy meadow where the adults can hunt for insects to feed their family. We spaced the four boxes going up the mountain through our network of meadows. This particular box, which we name "the cabin box" in a burst of noncreativity, is in clear view from the porch. It's about the size and shape of an old-fashioned camera, the kind photogs from the 1800s used, its entrance hole a cyclopean eye staring out on the world.

The two birds flutter in front of the box, inspecting a possible nest site. "That box will make a fine home," I whisper, hoping to convince them. But instead of flying in the entrance hole, they land on nearby branches. Then the male hops onto the edge of the entrance hole and *poof*, disappears inside. This is a good sign, I think. After a moment, he pops out again and flies up to the branch beside the female. Are they talking it over? How many places have they looked at already? Can't be fussing over the floor plan—it's just a one-room box! I expect her to check it out herself but instead they fly off.

Male bluebirds investigate several possible nest sites within their territory, leading their mate to the various choices, which she will inspect for whatever parameters a bluebird parent looks for. As with many couples, she gets the final say. But she is apparently not ready to buy. With my binoculars I watch them fly off around the bend of the L-shaped meadow, uphill to the north. I settle back in my chair with my coffee. Time will tell whether they decide to nest in our box.

Later that day bluebirds are again fluttering around the cabin box, but this time it is two males. Is one of them the male from earlier?

Suddenly I'm witnessing a fuss worthy of a reality TV free-for-all. The two fly at each other, fluttering and jabbing with their feet. They tumble to the ground in a flurry of feathers, but the fight continues. They jump at each other, flailing with their wings. One manages to get the better of the other, holding him to the ground like a schoolyard bully pinning a weaker kid, jabbing with his bill. Finally the vanquished bird scrambles away and escapes into the air. The victor sits a moment, then flies off.

My mind is full of questions I'll never be able to answer. Was this territorial battle between this morning's male and an interloper? If so, did he win? Or was he the one driven off? If he's the loser, will his mate abandon him for a more impressive male? If he's the winner, have they settled on our cabin box?

To my human eyes, bluebirds seem such pretty, gentle birds. But survival is about competing for resources. I've had a glimpse behind the curtain of bluebird life, seen beneath the sugary façade of a suburban cul-de-sac to discover seething rivalries and aggression—a bluebird version of *Desperate Housewives*. But I'm glad bluebirds are fighters. Over recent decades they've had to be, and it's not likely to change.

Bluebirds are cavity nesters, building their nests in holes typically in dead trees. With bills designed for grabbing bugs not excavating holes, they don't make their own but use tree cavities made by woodpeckers. But when people build farms and towns, cities and suburbs, they cut down trees. By the early twentieth century, all three bluebird species found in North America—western, eastern, and mountain—were (and still are) in a losing battle against human interests, their numbers plummeting due to loss of their habitat.

But bluebirds aren't drab, insignificant animals nobody cares about. People *like* bluebirds. In 1935, Dr. T. E. Musselman of Quincy, Illinois, came up with the idea of creating alternative housing for bluebirds by putting up artificial nest boxes. The occupancy rate of his first 102 nest boxes rivaled downtown Manhattan office real estate: 86 percent. The idea went viral, 1930s style. Audubon groups, local governments, and ordinary people all across the country who were horrified at the thought of a world without bluebirds got on board. Now trails of nest boxes extend along country roads and through public and private land across the United States and Canada.

Our four boxes are the beginning of *our* nest box trail.

———————

CABIN JOURNAL—JUNE 2000: *At first, all I could see inside the nest box was a pile of grayish feathers. Then I noticed a bright yellow V. Then another, and another. The nest box held five western bluebird hatchlings, piled atop each other like a litter of puppies. The Vs were their bright yellow bills. The newly hatched babies are naked and featherless, with stubs for wings, bright yellow beaks, and enormous, unopened eyes darkly visible through their pink, translucent skin. Pinkies.*

They react instinctively to the change in light inside the box, or perhaps the movement, as we open the box, raising their heads on unsteady necks, their beaks gaping wide. The vivid yellow of their beaks is a command the adult birds cannot refuse. If I were a bluebird, I would instantly stuff into that yellow target the grub I have brought in my beak.

By early summer of our first bluebird season, all four of our nest boxes are occupied, including the cabin box, though I will never know whether the residents are the bluebird pair I watched in the spring. Seeing hatchlings just hours old, I am in awe, startled at how fragile and vulnerable they are, these tiny packages of tissue-thin skin and air-filled bones no bigger or sturdier than a stalk of wheat. Their heads are enormous for their bodies, too heavy to hold up for longer than it takes to open their bills in a gape and accept the bugs their parents stuff in their gullets. Enormous eyes, still unopened, bulge beneath the skin. Bluebirds are visual hunters. Countless times we watch the adults peering down from a perch, or sometimes hovering, looking for insects in the grass. Those big eyes will be critical when these babies are on their own, hunting for insects in the grass.

The meadow is milling with bluebirds, mostly western, at least one mountain, flying about after each other like a tag team. We make a circuit to check the boxes, cautiously levering up the side access wall.

In the cabin box we find nearly helpless featherless pink hatchlings. In a box we have posted on a large, dead juniper snag (we dub this one "the snag box"), a female bluebird hunkers down on a grassy nest, motionless, her bright black eyes unblinking. She is sitting on eggs, and even our rude intrusion doesn't make her leave her young. We quickly close the box, as quietly as possible. In the box near the pond (yes, "the pond box") are five eggs colored a beautiful robin's-egg blue—no surprise, since bluebirds are members of the thrush family, cousins to robins.

In the fourth box, hung on a tall pine by the old campsite (you can guess the name . . .), we find a pile of big, wide-eyed babies with blue flight feathers, the remnant wisps of downy feathers glowing around their burgeoning adult plumage like a mist. They lie one on top of the other, so much larger than the hatchlings we just found in the cabin box that they are crammed inside their wooden home, transformed from helpless pinkies to feathered, bright-eyed birds. They are ready to fledge.

A day later, I go past the old campsite . . . and do a double take. The fourth box, the one that yesterday held bright-eyed young birds, is a broken pile of kindling on the ground. The brass wood screws that held it to the tree still shine brightly in the trunk but now bend downward ninety degrees. Tooth marks and deep scratches mar the wood of the box. A bear has pawed the box from the tree, clubbed it with a powerful swipe, then chewed on it trying to get inside. Successfully, it would seem. The remnants of a nest lie on the ground with a scattering of blue feathers. Those wide-eyed baby birds just gaining their flight feathers have been devoured by a bear.

There is a fluttering in the tree above me, something moving branch to branch. The adult birds, I realize, still coming to the nest box, confused and distressed. My heart hurts for them. I am a parent too. Their loss reminds me that I harbor forever the fear of losing my own child. Our lives are driven by the same desperate needs: to live, to survive, to raise our young, to listen to the rhythms of our unique existence and follow them.

I stand a minute working through my reactions. *Poor babies! Cool, a bear was here! Hey, we put up that box and you broke it!* In part, I'm angry at the bear for eating "our" baby bluebirds, though the bear is just following the rhythms of *its* role in the system.

Why do I react as if we have ownership of the young we help foster in our boxes? We are the interlopers here, arriving unbidden, making changes, putting up these artificial nest cavities to benefit birds. Now our efforts have secondarily benefited bears. The bear is not a villain. Like the bluebirds, it is a part of the natural community. How can I try to benefit one species while resenting another that takes advantage of my effort?

I have to laugh at my hubris. The bear visit was unexpected, but I should have expected it. A bear exploits any food source. Nothing in nature is static. Any change, any new development, produces actions and reactions. And a lot of them will be surprises.

We've had our first lesson.

CABIN JOURNAL—JUNE 2007: *A male bluebird is enthralled with the car and truck. He perches constantly on the side mirrors, flutters and attacks his reflection in the windows and windshield, doing noisy battle—bill-tapping and wing-fluttering—with his reflection in the car window. After we return from an errand to town, parking the truck and leaving the windows open, he flies inside and perches on the steering wheel and windowsill. Later an adult male (the same one as earlier?) perches on the very tip of the truck's antenna before flying up with his mate for a close look at the cabin eaves.*

Newly fledged young birds, still hanging together like nervous members of a teen gang, perch on the gutters, on the porch and deck railings, all over both vehicles. When one of the adults perches nearby, they mob it, squawking and fluttering their wings.

In our seventh year of fostering bluebirds, we realize a baby boom. Our bluebird trail, grown to ten boxes, winds up the series of meadows that rise westward up our property until the terrain becomes too rocky and wooded to be good habitat. We've named our two meadows. Elk Meadow, L-shaped, winds up from the campsite, goes across the slope where the cabin sits, then bends again uphill. Beyond a screen of trees and over a ridge lies Turkey Roost Meadow, named after we discovered a group of turkeys roosting in the tall ponderosas that border it. Rick makes a sketch to keep track of where the boxes are in these meadows. Though we've continued our creative box-naming—road box, lower Elk Meadow, upper Elk Meadow, power pole, Turkey Roost Meadow, salt block box (near a mineral block for the cattle that sometimes wander our land under a grazing lease)—he assigns each box a number to make record-keeping easier.

We try to monitor the nests without stressing the birds by opening the boxes too often. Adults flying in and out the hole is the obvious sign of an occupied box, but we often don't see that. The corners of the underside of each box are angle-cut to leave vents for air flow. Instead of opening every box, we start by peeking underneath. If there is a nest inside, we can see grass and material poking out through the corners.

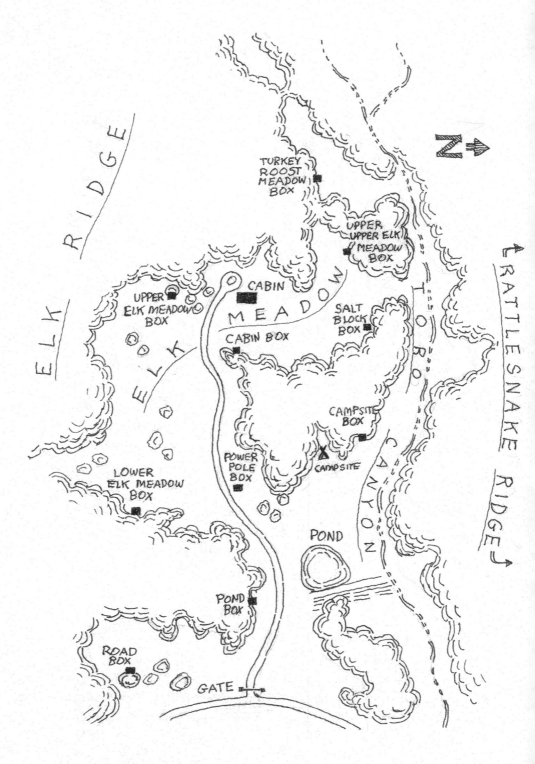

All the boxes are occupied, and the meadows are busy with blue-birds hunting insects. This year's young from the cabin box, grown into speckle-breasted babies, have left the nest to perch on branches near the box. Now they pester their parents like lazy teenagers.

Baby: (*Bill gaping wide open, wings fluttering.*) "Feed me!"

Adult: (*On branch near baby.*) "Food is in the grass right here. Watch closely." (*Drops to ground, grabs grasshopper, returns to branch, stuffs in baby's beak.*)

Baby: (*Gulps, bill gapes open, wings flutter.*) "Feed me!"

Adult: (*Bluebird version of exasperated-parent sigh.*) "Kid, I told you, food is in the grass right here." (*Drops to ground again, waits, but baby doesn't budge, finally grabs bug and returns to branch but doesn't feed immediately.*)

Baby: (*Bill gapes wide, goes into paroxysms of fluttering and feather trembling.*) "Feed me!"

Adult: (*Stuffs insect in baby's beak, flies off.*)

Baby: (*Fluttering, squawking.*)

Adult: (*Lands back on branch but no insect.*)

Baby: (*Paroxysms of fluttering and gaping and squawking.*)

Adult: (*Flies off.*)

Baby: "Wha-at?!" (*Looks around perplexed. Squawks. Flies off.*)

We haven't checked the pond box since a month ago, when it held five blue eggs. Long hatched, the young should be pretty much ready to fledge, if they are even still in the box.

Rick, Olivia, who is now nine years old, and I approach quietly. Adult bluebirds fly around the meadow, others perch on the power line, but none flies into this particular tree. I step close to the box, pull the nail that holds the side access wall closed, and carefully tilt the side wall up. I just have time for a glimpse of a grassy nest, pounded nearly flat by all the feathered babies crammed inside, when one of those babies leaps for daylight straight past my head!

I quickly close the box, Rick grabs the dogs and pulls them away, and Olivia spots the baby bird—looking surprised and uncertain—in the grass. Do we return it to the box to let it fledge on its own time frame, or just let things be?

"We have to save it, Mommy!" Olivia says.

"OK," I nod. "Let's put it back in." She bends to capture it, but the baby flutters awkwardly away through the grass. So much for our role as benign, noninterfering landlords. The two of us stumble through clumps of grass, swerve around prickly pear, trying to corral the baby, which flutters awkwardly just beyond our reach. This situation is getting out of hand! But it doesn't take flight, which tells me maybe it needs more time in the nest.

Finally Olivia gently scoops it up. With the young bird blinking, bright-eyed, in her cupped hands, she looks up and smiles. "She's all trembly. I can feel her little heart beating. I think she's scared." She turns determinedly back toward the nest box. I don't need to tell her to carry the baby gently; she is resolute in her guardianship. Carefully we put the young bird back into the box through the entrance hole.

But what a fine mess we have stirred up. The parents now flutter about the tree, agitated at these nest marauders. They fly at us, trying to drive us off. We back off to a safe distance to watch as one of the parents flies into the box to check on its young. Apparently satisfied that all is well, it flies back out the hole. In the next second, a fledgling follows! It flutters tentatively into the neighboring piñon with the "How do you drive this thing?!" uncertainty of a bird on its first flight. The head of a second fledgling now pokes out the nest hole. Then it, too, launches, fluttering to a branch near the first.

"I think those babies wanted to get out of the box," Olivia says, her face solemn, her tone reproachful. "But Mommy, we should be more careful when we check the boxes."

"I think you're right." I put my arm around her as we continue our walk, glad my wise little pinkie still has many years before she will be ready to fledge.

Babies are everywhere. Each spring and summer, the land bustles with life. Tiny chorus frogs, no bigger around than a nickel, blossom at the edge of our pond, newly transformed from water-bound tadpoles. Inch-long prairie lizards dart around the rocks, tiny replicas of the adults that sun themselves on our kitchen step. Families of wild turkeys forage in the meadow, the adults leading broods of up to ten tawny poults that mill around them like curious preschoolers. Long-legged mule deer fawns, white spots gleaming on handsome brown coats, appear from the edge of the trees, wide-eyed and curious as they follow their cautious mothers.

Hiking a trail through the oakbrush one day, we come nose to nose with an adult coyote. For one endless second we all freeze—four-legged and two-leggeds—then the coyote breaks the spell, reverses, and disappears. We grab our dogs and retreat back down the trail. That explains the noisy family "sings" we've heard every evening from back up the arroyo, a coyote family sounding off before heading out to hunt. I hope the pups are not so young that they're still in the den and that our intrusion doesn't lead the adults to move them. In the fall, we find their old den—a large, well-used burrow beneath a rocky overhang, the bones of small mammals scattered around the pounded soil at the entrance.

We are part of a community of families, our daughter, Olivia, our little human "pinkie." When she finds a newborn pinyon mouse—as pink and unformed as any newly hatched bluebird—lying abandoned on a pile of lumber by the old campsite, her first question is, "Can we keep her?" As earnestly as she insisted that we put the escaped fledgling back in the nest box, Olivia wants to rescue this helpless mouse. And though I know this may not end well, I can't leave this baby here either.

She names the mouse baby Luna. We make Luna a soft nest of old cloth in a small box. From the back of the what-not drawer, I fish out a miniature baby bottle, complete with nipple, specifically meant for tiny orphaned animal babies, an item that I'd gotten I can't remember where.

We warm some milk and put an ounce or two in the bottle. Holding Luna carefully in a soft cloth, Olivia coaxes her tiny mammalian "pinkie" to drink, smiling in wonder as the little mouse actually sucks at the bottle, eyes still tightly closed. We bring Luna back to Denver with us and she grows, but even after ten days her eyes aren't open and she doesn't move around much.

I fear there is a hard lesson waiting for Olivia. One morning her precious Luna lies stiff and motionless in the box. As a biologist I try to explain her death—she was probably never healthy, which is why the mother set her out of the nest where we found her. She was probably never going to survive.

But Olivia is devastated by the loss of Luna. We have often found the remnants of dead wildlife on the land, from mule deer carcasses cached beneath leaves by a mountain lion to occasional dead nestlings in the nest boxes. But Luna was Olivia's baby. And so she learns the hard lesson of love and loss that is a part of life. And like her mother, she turns to words to help come to terms with it. "Luna, What You Were to Me" she titles her poem.

"Oh Luna, how I miss you," she writes, "Why did you have to go?"

We walk through a moonless night down the drive toward our old campsite. The dogs, Jasper and Rosie, range quietly around us, investigating the world of scent clues released by the cooling evening. The sky is a blizzard of stars pinned against black onyx, cold and infinite, reassuring in its constant presence. The Milky Way lights a glowing path I want to dance along. Where would it take me? In the southern sky, Sagittarius the centaur follows the dagger-tailed Scorpius, with Antares its bright heart, in a slow and constant chase. In some few months Orion the hunter will rise to guard our winter nights.

The stars illuminate the night enough that we make our way without lights, preserving our night vision. We pass through a wondrous mist of sweet fragrance, stop to breathe in great gulps. Night-blooming scarlet gaura releases its perfume in the cool darkness, beckoning moths and other nighttime pollinators. Freeloaders, we share in its offering.

Beyond the pond I see the shape of a nest box on a piñon. I think of the bluebirds, sheltering now in the dark security of their nest boxes, mothers nestled down on newly built nests, incubating eggs the color of the sky. Protected within those eggs are the seeds of birds, begun as no more than a collection of cells but growing, growing, beneath their mother's warmth. Soon those eggs will ripen, the birds grown so large they lie curled tightly within the shell, restless to unfurl and enter the world. The need to be free, to be born, will drive them to crack open the shell from within. Just a tap to make the first crack, then another and another, *pip, pip*, until the tip of their beak pokes into the air for the first time. Pippin. That's the quaint name farmers call chicks pecking their way from the shell. "Pippin," I say softly to myself, enjoying the sound of it.

It is for this the mother bird waits, nestled in her dark box. Will she hear that first pip? Will she rejoice? When her pippins have freed themselves, she will nurture them from pink babies to feathered fledglings. Watch them launch. Will she lament their leaving, as I will when my young one launches?

If it's early enough in the season, she and her mate will lay a second clutch, and sometimes, a third. Migrate for the winter, return in the spring. Nest and incubate and brood and launch more young, who will move also through the cycle as countless generations have before.

It's the natural rhythm that all life on Earth is a part of, including us. The natural world proceeds in rhythms, in circles, that vary year to year but add up over many years to balance. Some are bad years—too dry, too wet, late spring storms—which are balanced by good years. Followed by bad ones, made up for again by good. No beginning or end, an endless hoop. Things change and the circle adapts, rebalances. But major change that happens too rapidly doesn't allow the circle to adapt. And the circle dies away.

Nature did a lot of rebalancing in 2001, just the second season of our bluebird trail. That was the year we call the Summer of the Bear.

4

THE SUMMER
OF THE BEAR

Warmer temperatures and use of [human] foods . . . reduced
black bear hibernation, suggesting that future changes in
climate and land use may further alter bear behaviour and
increase the length of their active season.
—"Human Development and Climate Affect Hibernation
  in a Large Carnivore with Implications
  for Human-Carnivore Conflicts,"
  Journal of Applied Ecology, September 19, 2017

CABIN JOURNAL—JUNE 2001: *Tuesday night, as we sat on the deck listening to a poorwill calling from the dark, we heard a loud* crack! *from the trees beyond the meadow. Then, a double* crack! crack! *We peered into the blue-black night, wondering what was out there, and what it was up to. In the morning we discovered the answer.*

The American black bear, *Ursus americanus*, has an olfactory ability seven times that of a bloodhound. Bears are probably the keenest smellers in the world. A bear can detect food submerged in water. It can smell food inside a cooler. It can pick up molecules of food odor on clothing that has been carried in a pack along with food. It can detect the scent of a human who passed half a day earlier.

I should have remembered that fact.

———————

January 1. The end of the first day of the New Year, an hour after sunset. We stand on the deck looking to the southwest. Venus and a sliver of moon pose in conjunction, gazing at each other like lovers. It is the bitter season, the air hard and brilliant as jagged ice. There are no sounds of nightbirds, no arias from crickets. Only the bright and silent burn of stars, points of cold light sparking against a void as black as a bear's pelt.

The bluebirds have gone for the winter to the relatively moderate climate of southeastern Colorado. It's the lowlands, by Colorado standards, some twenty-five hundred feet lower in elevation than here, a land of grass and sky and unexpected canyons that cut the prairie. A place not greatly different from our land, though with fewer trees on the uplands. Like Swedish immigrants who came to America and took up residence in Minnesota, in a climate and terrain that reminded them of home, our bluebirds migrate for winter to a landscape not unlike this one. Do they dream, I wonder, through long winter nights of the summer days and blue evenings of this piece of the wild?

By spring equinox, the bluebirds will be back, searching for nest sites to raise their young.

———————

Early April. Elk tracks emboss the muddy ground like cuneiform pressed into clay. On a day of crystal sky, we follow their tumbled trail up Elk Ridge and climb the series of pine-covered saddles that rise like steps toward Montenegro. This seventy-six-hundred-foot ridge rises behind our cabin, the highest point in the area. From the top we can see the Sangre de Cristo Range—the southern tail of the Rocky Mountains—marking the western horizon twenty-five miles away. Among these peaks lie ten that rise fourteen thousand feet or higher in elevation. To the northwest, also twenty-five miles distant, lie the Spanish Peaks, so that the point of Montenegro forms the narrow-angled top of an isosceles triangle with equal arms to each of these landmarks.

The world is awakening, and I sense that the seasonal movement of creatures from winter haunts to summer ones is on, animals shifting in predictable directions like clouds moved by prevailing winds. In Colorado, wildlife follows the bloom upslope in spring and summer. Soon the elk will move to the high country for summer. And the bluebirds will return.

Where a swath of ponderosas tumbles down a north-facing slope into a side canyon, we find three pine saplings that have been rubbed by bears. One tree is bent double, like a tired old man, or maybe a croquet wicket. If giants were the ones playing croquet.

The tree's bark is worn off in a ragged patch on one side, fresh beads of pine sap oozed into the wound, sticky and fragrant. I look close, find coarse black hairs caught in the sap, a few finer, blond hairs. I imagine the bear, newly emerged from hibernation, still groggy from its long sleep but desperate to scratch. Tickled by the woolen underwear of its underfur, it ambles up to the sapling, tests it for spring with a bulky paw, then begins to rub. First, perhaps, the side of its neck, then its chin, its muzzle, the top of the head. Then it turns around to scratch its posterior against the tree, up and down, doing deep knee bends with its hind legs. Then it maneuvers so the tree scratches its back, eyes closed in feels-so-good ecstasy, scratching that winter itch stored up over months of cold weather sleep. The bear leans backward more and more, the sapling bends with its weight until *whoa!*, the bear over-tips and rolls to the ground. Undaunted, it begins to rub its belly against the sapling, clambering over the young tree until the trunk is permanently bent. There are older trees on our land, still living, but shaped permanently into croquet wickets by itchy bears on long-ago spring days.

Biologists debate whether the main purpose of tree-rubbing is scratching an itch or scent-marking trees to alert other bears. In some parts of the country, rub trees are large, sturdy snags used year after year. But on our land, we find only springy saplings rubbed, and new ones every year.

Like greening grass, bear-rub trees are a sure sign of spring.

———————

Mid-April. Bluebirds have arrived. We walk past the splintered remains of the campsite box, destroyed by a bear last year, still lying forlornly at the base of the tree. Checking the three surviving nest boxes, we find the cabin and pond boxes empty, but the snag box has a few strips of juniper bark, the beginnings of nesting material. We have brought four new boxes and use deck screws to hang them on trees. One we place in Turkey Roost Meadow. Another we put at the lower end of Elk Meadow. The third box we place on a ponderosa at the front of our lot not far from the road, referring to it as "road box." The last we hang on the tree to replace the one earlier destroyed by a bear, renaming it "bear tree box." Now we truly have a bluebird trail—a line of nest boxes extending across our land.

By Sunday everyone has gotten the memo, and there is an explosion of bluebird activity. They hunt insects in the meadows, fly to the nest boxes, peer in the entrance holes like prospective tenants inspecting the number of bedrooms. We hear them chirping and calling, flying back and forth in groups of two, three, and four. Some take their rest on the power line, sitting in rows like little leaguers on the dugout bench.

In the mud alongside the driveway, just across from the kitchen door, I find a set of four-toed paw tracks. They are wider than they are long, rounded at the toe, and don't show any claw though they are set deep into soft soil. I measure them—four and a quarter inches wide and three and a half inches long. I don't need to consult a tracks guide to know who left them. A mountain lion. We resolve to not let Olivia play outdoors more than a few feet from us.

Memorial Day. The scream of a red-tailed hawk sounds above the meadow, a shrill, descending *keee-eer!* tumbling from the sky. Two hawks soar in unison over Elk Ridge, swirling around each other like flirting teenagers.

The meadows are alive with bird activity, as if the circus has come to town. Ravens chase each other through the sky, their shadows sweeping the grass like scythes. Turkey vultures soar upward on warm air currents then glide back, surfing the waves. Violet-green swallows swoop jetlike along the eaves of the cabin. Bluebirds fly continually between meadow and boxes. A late-nesting pair carries strands of grass in their bills, just beginning to build their nest when many of their contemporaries are already on eggs, or perhaps already feeding hatchlings.

Hummingbirds do battle around the feeders. As yet, few flowers are blooming and the hummers flock to the sugar water we set out for them—one part sugar to four parts water, no dye required. If I unknowingly step in their flight path along the porch, they nearly part my hair.

One afternoon, returning in the car from a visit with neighbors, I spot a dark shape near the cabin. A bear, and a good-sized one. It doesn't seem to hear the car approaching. Perhaps the low hum of the engine doesn't rise above the sigh of the wind. As if in slow motion, the bear rises slightly on its hind legs, front legs dangling, lifts its nose to the cylinder of the hummingbird feeder that hangs near the north window.

This 250-pound Winnie the Pooh is looking for honey—the sugar water in our feeders. We are meticulous in taking in our feeders at night to avoid ever rewarding any bear with food, but we have not worried about them during the day. Rick honks repeatedly and the bear hesitates, frozen for a moment. It may not recognize a car horn but surely the sound frightens it. I get a good look at the bear, the heavy head with rounded ears, the doglike muzzle ending in the snuffling nose of a forager. Its coat is a glossy black. Then it turns away and lumbers into the trees behind the cabin in a rolling gait.

The black bear is a large animal, equipped with fearsome teeth and claws, but it has an undeserved reputation as a ferocious predator. Anyone who has seen a bear run knows it is not an animal designed to chase fleet-footed prey. Up to 90 percent of the black bear's diet is plant material. The meat it consumes is mainly rodents, insects, and carrion.

Food. It defines the character, behavior, and activities of a black bear. Bears are in a constant search for food, a constant struggle for survival. They have a lot of bulk to sustain on berries, nuts, and grubs. When bears emerge from hibernation in spring, having not eaten since

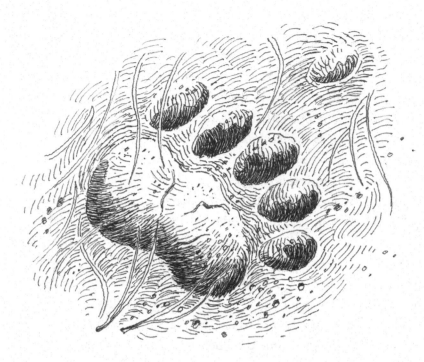

fall, they are thin, their body fat having metabolized to keep them alive through long months in the den.

By late summer, the bear is in an even more desperate search for calories, racing the seasonal clock. Autumn is approaching, and the life-threatening months of winter. If a bear does not put on sufficient body weight to sustain its hibernating bulk through winter, it will die. Hibernation places even more demands on female bears. Many begin their winter sleep pregnant with one to three young. Their embryo-like cubs will be born in the quiet darkness of the hibernation den. Mama bears must have sufficient body fat to sustain themselves and nurse their babies. If the mother is too thin when she enters hibernation, the embryos may not implant in the uterus, and she has a chance to survive winter. If the young do grow but her fat store is not sufficient, the cubs, and even the mother, may die.

At birth, each cub weighs about half a pound, the same as a cup of water, and looks more like a rodent than a creature destined to become a mighty *Ursus*. The mother cuddles her tiny babies to her furry bosom, suckling them and keeping them warm, until they emerge with her in spring, the size and shape of large St. Bernard puppies. Mama bear must now find food for herself and her fuzzy, roly-poly cubs. They are still nursing, but she begins to teach them how to be bears—what to eat and where to find it. They will den with her again in the fall, half-grown ursines that still need their mother. But emerging from the den as yearlings, they will face a cold, hard world. Mother bear drives them away and they must begin life as independent adults.

With the grim face of winter peering in the window, late summer through fall is a time of desperation for bears. They spend up to twenty-three hours a day foraging and feeding. This is when bears are most likely to get into trouble with people. When people living in bear country leave their trash and barbecue grills out, they lure hungry bears into what can be a death sentence. Colorado has a "two strikes and you're out" rule for problem bears. State wildlife officers will trap and relocate a bear once. On the second offense, the bear is destroyed. Bears don't understand this rule, of course. All they know is that they're hungry. In lean years, a bear's options are grim—a slow death by starvation or, if it goes seeking food among people, a quick one from the bullet of a wildlife officer. Shooting a bear they have caught in a trap is a duty the wildlife officers hate with a passion.

One fall several years earlier, we saw a bear in a crab apple tree along a street in a mountain town. It was frantically gobbling the apples even though surrounded by a crowd of gawking, photo-snapping humans. Only a desperate hunger would drive a shy animal to reveal itself and keep feeding within view of its most dangerous predator.

Early June. I notice birds checking out the cabin nest box, but they aren't bluebirds. They are slender-bodied and long-tailed, with "bumped-up" heads, whitish throats, and washes of yellow across their bellies. From a perch in a nearby piñon, one flies out, grabs an insect in flight, then loops back to the same perch. It is fly-catching, a hunting technique typical of a group of birds called—surprise—flycatchers. The birds inspecting our box are ash-throated flycatchers.

Throughout the day, the flycatchers mill around the box. Finally one enters. Will they adopt this box as a home?

A week later, I check our bluebird trail for occupants.

Turkey roost box: five pale blue eggs—bluebirds

Snag box: five blue eggs—bluebirds

Cabin box: four white eggs with jagged brown streaks striping the egg longways—ash-throated flycatchers

Bear-tree box: five blue eggs—bluebirds

Elk meadow box: no activity, a few strands of grass

Pond box: a mother bluebird on nest

Road box: no activity

———————

Mid-June. The birds are engrossed in parental duties. From the cabin we have a perfect view of the cabin nest box. The ash-throated flycatchers are very busy flying back and forth with insects in their bills. They are feeding young. Bluebirds do the same at the snag nest box. Western bluebirds and ash-throated flycatchers incubate eggs for about fourteen days; the flycatchers then brood their hatchlings for two more weeks, and the bluebirds for up to three.

It is the time of the summer solstice, the longest days of the year and midsummer, when magical things can happen. We sit on the deck enjoying the blue evening. The sun sets behind Elk Ridge but the light lingers, seeming to enjoy the summer evening as much as I do.

Finally, darkness. The plaintive call of a poorwill hunting low above the grass comes to us from the meadow—*poo-wee-ee, poo-wee-ee.* "Poor me" it seems to say. We listen as the call of this night bird fades to the south.

Suddenly from the dark comes a loud *crack!* We move to the deck railing, straining to see something, anything, in the inky night. But our human eyes are designed for daytime vision and we are largely blind once the sun has closed up shop. We stand, curious, wondering what is abroad in the night in our meadow . . . its meadow . . . bold enough to make such a racket.

Then the sound comes again—*crack, crack!* This is no night bird, no stealthy, nocturnal hunter. We stand a long time, listening to the night, until our eardrums throb from trying so hard, but we hear nothing more.

Next morning, I am eager to find the source of the sound. A brief search reveals it. The bear tree nest box has been cleaved open as if someone took an axe and struck it a mighty blow. Split like a watermelon, the box still hangs from the tree. Its two halves splay in a *V*, like arms raised in supplication. Only one animal could have done this. A bear.

A friend once told me of a buddy who kept a two-hundred-pound chest freezer, locked, on the screened porch of his mountain cabin. A bear broke into the porch and rolled the freezer like a bowling ball out the door, down the steps, and one hundred feet down the mountainside.

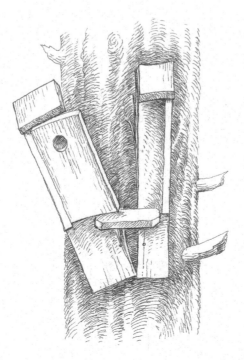

At the bottom of the hill, the locked freezer capitulated, and the bear pried it open and ate all the steaks inside. And the ice cream.

I imagined the bear afterward, lying in the shade of a tree, eyes drowsy, belly swollen from the feast of steaks now thawing inside, its muzzle wearing an ice cream mustache.

If a two-hundred-pound steel freezer is not a challenge for a bear, a wooden box of three-quarter-inch pine is no problem.

On the ground below the broken box lies a nest, a neatly woven bowl of long grasses with a few blue feathers worked among them. At Memorial Day, that nest held five eggs, but there is no evidence of any surviving young.

Below the wounded box, the stump of a newly broken branch oozes sap. This must have been the first *crack* as the bear tried to climb up on this branch. There are teeth marks gouged in the soft pine of the box. I imagine the bear trying to gnaw its way into this box that held a meal, then, frustrated, whacking it a good one. Whacking did the trick.

A pair of bluebirds moves about the branches above us, from one vantage point to the next, then back again. As a biologist, I'm not supposed to attribute human emotions to animals. But is it so outlandish to think the adult birds are distraught at the loss of their young, bewildered at the destruction of their nest?

We pull down the remains of the box. It is not repairable. I remember all I have read about the memory of a bear, how it never forgets a food source. A bear that has gotten food from a picnic cooler will peer through a car window, spy an ice chest, and break into the car to get it. Don't think locking the car door will help. A bear does not need a Slim Jim. It pries the top of the door away from the frame, then tugs, sits on, and pries until it tears the door from the car.

I ask myself which is the more intelligent species—the bear that remembered getting food from this tree last year or the people who put up a replacement box at that same spot?

———————

I'm angry. It is two days since we heard the *crack* in the night and the marauder has struck again. We find most of the lid ripped off the box in Turkey Roost Meadow. The box itself still hangs on the tree, otherwise intact. Inside the now open-air box huddle five feathered baby bluebirds, their black eyes bright.

What would it have been like to be those babies, trapped in the nest while a monster gnawed, pawed, and swatted at your home, trying to get you? Like feathered three little pigs, they huddled indoors as the big bad bear huffed and puffed and blew the roof off.

I imagine the bear thrusting its nose into the box to gobble up the babies, but once its boxy muzzle is wedged inside, it can't open its mouth. Like a man who inserts his fist into a pickle jar but then is unable to open his hand to grab a pickle.

The baby bluebirds seem undamaged. They wear the stiff flight feathers of mature young. They are ready to test the winds, should fledge at any time. Surely, I think, the fates have saved them for great things.

I wonder whether we are being watched. The bear, I remember, is smarter than we are. *Leave our boxes alone!* It was exciting at first, having a bear around, but now . . .

Beneath some ponderosas in the woods below the cabin, we find two daybeds—piles of pine needles pushed up to make a nest and packed down in the middle by the weight of a large body. Has the bear been napping here, as we went about life just up the hill and across the meadow in our cabin?

We discover more and more evidence of our ursine neighbor: round piles of dark bear scat peppered with pine nut shells, chokecherry pits, and the undigested berries of three-leaf sumac; rotten logs torn open and scraped for grubs; large boulders rolled over by something much stronger than either of us to get at the grubs and insects beneath. The blunt-cut stems of yellow sweet clover, white peavine, and purple penstemon stand forlornly, their blossoms nipped off. I'm reminded of Morticia Addams from *The Addams Family* TV show, who prepared long-stemmed roses for the vase by clipping off the blooms and arranging the stems.

We check the roofless box in Turkey Roost Meadow. The babies are still there. *You gotta get going! Now!* I want to give them a send-off party and launch them into life, but they must fledge on their own schedule.

Returning to the cabin we discover that the snag box that we passed earlier in the day now lies shattered on the ground—*whacked.* There has been a third attack, just fifty yards from the cabin! Is this some ghost bear that moves around us at will, never showing itself, but taking advantage of the food we have provided, neatly packaged in wooden boxes?

That evening, we check the pond box. As I carefully lever up the side, a young bluebird tumbles out. I shut the box and manage to delicately gather the baby bird from the grass. Cupping it in both hands, I show it to three-year-old Olivia. Her eyes are huge, entranced to see this fragile, wild baby so close, but she has been among wild creatures her entire short life and knows not to squeal or grab for it. She reaches out one small finger and strokes the baby's wing oh-so-gently. Then carefully I slip the fledgling back into the box through the entrance hole.

The poor parents flutter frantically around us, and we retreat. The female enters the box to check on her babies. As one mom to another, I hope she is reassured.

Walking back up the driveway, we notice deep scratches spiraling down one of the power poles, at a height of about four feet. I put my fingers to the scratches. Five parallel grooves rake down both sides of the pole, left by the five claws of a bear's paws as it clasped the pole and tried to climb. Ten feet above my head is a wooden box we put up several summers ago as a roosting site for bats. The bear has made the association: wood box equals food.

Our ghost bear is one smart bruin.

The next morning, we take a walk, making a circuit of the nest boxes. The roofless box is empty. The babies have fledged. We notice a number of rub trees and fresh bear scat in several places among the meadows and down the arroyo.

Unable to keep away, I go out again about 4:00 PM with Jasper, my big retriever, making the circuit of the boxes. I wonder whether I should worry about encountering the bear. I've never before felt afraid outdoors in Colorado—I fear people but not nature. Black bears are shy animals, much more likely to run than endure a confrontation. Attacks on humans by black bears are rare, but I'm glad Jasper is with me.

As I come up from the arroyo into Turkey Roost Meadow, I see the damage. The bear has been here again and this time succeeded in tearing the roofless nest box from the tree, ripping it open like a bag of potato chips. But it got no snack. The potato chips fledged this morning.

Finally I get it. The bear has been following our scent. The bear learned that we make a circuit of the boxes. It might have been watching us, but it didn't need to. This bear has a nose, is the best smeller in nature. It can detect the scent of a human who passed by half a day earlier.

We have been leading the bear to the boxes.

June is a fine month in southern Colorado. Sideoats grama and western wheatgrass are at their tallest. Scarlet globe mallow and prickly pear are still in color. Purple penstemon, Indian paintbrush, and other summer bloomers have begun.

I sit on the deck, coffee in hand, scanning the meadows around the cabin. Hummingbirds constantly whistle past, coming in to feed, then moving off for a twenty-minute rest to digest the fraction of a teaspoon of nectar filling their tiny stomachs.

I notice a large animal near the nest box at the lower edge of Elk Meadow. "Looks like an elk," I say to Rick, noting the buff-colored body and dark legs. But it doesn't seem big enough. Then the animal rises slightly on its hind legs. I grab the binoculars.

It's a bear. *The* bear? Its back and sides are blond—black bears can be black, brown, cinnamon, or even blond—but its legs are darker. Can this be the same stout, glossy black bear that sniffed our hummingbird feeder at Memorial Day?

I watch as the bear presses its snuffling nose to the entrance hole of the nest box. This box is empty, I already know. The bear probably does too, but it has learned that wood boxes mean food. Maybe this time it will contain a meal? Not that baby bluebirds are much more than a mouthful for a bear.

The bear does not bother the box. It drops back to all fours, lumbers along the edge of the trees, heading north. At the driveway it stops, swings its head our direction, and I imagine that, for a moment, the bear and I make eye contact. The bear is medium sized but lean. Its coat is rough, scruffy, not the smooth coat of a well-fed animal. Suddenly I am ashamed of my anger at this hungry animal's destruction of our nest boxes. We can buy more for just a few dollars, put them up for our own amusement. But the bear faces starvation every day. *You can have the bluebirds!* I think, as if they were mine to give.

I am caught by the dark gaze of this magical creature who dies in fall and is reborn each spring. I want to hold on, for just a moment, to this glimpse of wildness, to hear through it the heartbeat of the land, to feel a part of the natural world in a closeness my species gave up long ago.

But the bear knows me for what I am, the most dangerous animal on earth. It turns and lopes across the driveway, then across the meadow by the pond, disappearing into the trees.

Bears are the kind of neighbors you rarely see but know are around because you notice their patio furniture has been rearranged, you glimpse their garage door going down, you see their recycling left at the curb on alternating Mondays. Or you discover that they have, once again, busted down a nest box.

Bears were inaugural members of our nature journal, making an appearance in the very first entry: *1995—Fall Sightings: black bear—lots of scat, bear rub tree with coarse black hairs in sap.* They leave their calling cards every year, not just scat and rub trees but also daybeds, overturned boulders, tracks in soft soil. We've seen bears many times, and every year we lose one or more nest boxes to them, though no other year has had bear activity as dramatic as the Summer of the Bear.

CABIN JOURNAL—AUGUST 21, 1999: *First actual bear sightings! Saw a bear (large, almost black) on the road just at the Toro Canyon road wash. Later found tracks of a female and cub in muddy area by the road.*

A bear family is in our area. Saw a female with three cubs at midday, from the cabin porch! Good-sized female, dark brown. We've seen an incredibly large number of scat piles in the deep drainage by the big ponderosas up our arroyo. We've come to call it Bear Alley. We found six distinct scat piles in one small area of the arroyo. Driving out Sunday PM we saw a smallish dark brown bear in Long Canyon.

AUGUST 5, 2000: *Spotted some very clear bear tracks behind the cabin. The bear had been walking south to north in the muddy ditch just west of the cistern. Front track = four and a half inches wide by four inches long; hind track = five and a half inches wide by six inches long. I made a plaster cast of the tracks.*

AUGUST 8, 2009: *In the distant meadow visible across Toro Canyon we see a large shape—a cinnamon bear with darker legs ambling across the open ground. Then behind her comes a small, round, brown shape—a cub. They move out of the clearing out of sight to the left, then suddenly, a small, round black shape gallops across the clearing after them—a second cub. The bears are in the clearing less than a minute. I wonder if the mother is reluctant to be out in the open for long, not taking time to forage.*

JULY 10, 2010: *By the pond up the road by the hunting camp, saw a female black bear with two cubs. They were up in a tree, and we saw her move into the forest with one cub following and a few moments later saw the second cub drop from the branches and gallop into the woods after its mother. All bears very black.*

But the journal reflects a change over the years. Black bears seem to be active later into the fall. It's hard to quantify—the journal is anecdotal and the entries intermittent. And dates alone don't prove anything. Warmer weather and abundant fall food have always kept bears out and about later. In Colorado, depending on conditions, bears will start hibernating any time from late September to December, with peak denning time being late October.

Still I wonder, are the hibernation schedules of our bears changing?

THANKSGIVING 2010: *Signs of recent bear activity visible—several fresh bear poops along the driveway in Elk Meadow.*

DECEMBER 30, 2010: *LOTS of bear sign. We appear to have an out-of-hibernation bear that has been active here since Thanksgiving. He/she deposited a pile of scat directly in front of the door to our shed some time in the last month. Then we found abundant bear poops in and along the north fork of the arroyo, just up from the old Bear Alley. Some piles were <u>huge</u>. Some were fairly fresh. We're guessing the mild, dry weather has delayed this bear from entering its hibernation den. This is probably the same animal that busted our bird feeder prior to Thanksgiving.*

NOVEMBER 10, 2017: *Sunny and sixty-four degrees at noon. Signs of bear activity: two large, fresh bear poop piles high up on the slope of Montenegro.*

OCTOBER 25, 2020: *Door busted off pond nest box since our last entry on October 4.*

JANUARY 18, 2022: *Lots of bear scat around the fire ring by the cabin, farther up by the gas well to the northwest, on the flat top of the Thumb, and in various of the higher meadows above Turkey Roost Meadow.*

I don't remember seeing all this bear scat when we were last at the cabin after Thanksgiving, when I wrote in the journal *Extremely dry and warm—sixty-eight degrees. Very sunny.* Was it left in the eight weeks since we were last here? Another bear or bears active in December?

Bear scat is not a rare commodity on our land (if only I had a free cappuccino for every one I've seen) and rarely rates a mention in the journal. We recorded the November, December, and January sightings because the timing was so unusual.

A five-year study by Colorado Parks and Wildlife in the mid-2010s found that warmer temperatures were keeping bears out of hibernation until later in the fall. As long as there was food around, they kept active and eating. If that food came from humans, in the form of mishandled garbage, birdseed, or landscaping plants, the bears were likely to get in trouble and end up dead. "We speculate that longer active periods for bears will result in subsequent increases in human–bear conflicts and human-caused bear mortalities," concluded the CPW study ("Human Development and Climate Affect Hibernation in a Large Carnivore with Implications for Human-Carnivore Conflicts," *Journal of Applied Ecology*, 2017).

When bears hibernate, a whole range of magical changes happens. Their digestive system goes on standby. They don't eat or drink or defecate. Their breathing and heart rates reduce, though their body temperature drops only a few degrees. When they emerge, it takes a while for the system to rev up again. A spring bear may be out of the den and active for a few weeks without eating or relieving. This zombie-esque state is known as walking hibernation. It's why we find newly rubbed saplings and other bear sign in the spring but no fresh scat.

APRIL 17, 2013: *Rick finds a bear front paw track in soft soil by the cistern. In spite of my intentionally looking, no bear scat seen anywhere.*

Sometimes a hibernating bear will rouse, and if it's warm enough, may briefly leave the den. But the bear's system is still hibernating—it isn't eating or pooping. That tells me the bear or bears who left us fresh calling cards in November and December had not begun hibernating yet.

So what happens to bears as the climate gets warmer? Their metabolism—which is a bear's calendar—will tell them it's still warm, they don't need to hibernate yet. But natural foods may be disappearing, not from cold and winter but from higher temps and drought. Bears can, and will, alter their behavior, but they can't alter their biology or their need for food.

It's a worrisome future for this magical animal, whose ability to die and be reborn may be of little help in a warming world.

5

AN AUTUMN FOR ELK

Climate change in the form of reduced snowfall in mountains is causing powerful and cascading shifts in mountainous plant and bird communities through the increased ability of elk to stay at high elevations over winter and consume plants, according to a groundbreaking study [from US Geological Survey and University of Montana].
—ScienceDaily, January 11, 2012

CABIN JOURNAL—OCTOBER 1999: *9:00 PM—from the dark meadow we hear a high-pitched moaning. Then bleating and mewing like souls lost in the night. Suddenly nearby, from the south, a distinct elk bugle.*

We turn off the cabin lights and quietly, gently, ease open the door onto the porch. The autumn night pulses with music. Coyotes perform from near, in the arroyo, and far, down Long Canyon, as the song-dogs gather for their nightly sing. A great horned owl hoots from the tall ponderosas below Rattlesnake Ridge, where moisture gathering at the base of the slope nurtures their growth. Farther off, to the east, comes the whinny of a western screech owl, accelerating and dropping in pitch like a ball bouncing down a flight of stairs. And from our meadow, those high-pitched moans that perplex us.

A waxing half-moon brightens the night, and through binoculars we make out the movement of phantoms in our meadow. Large shapes, small shapes. Elk. We estimate forty to fifty animals. The cows call to their calves with mews and bleats. To the east, to the west, the bugles of bull elk, starting low then rising to a high-pitched screech, tapering off to *chuffs* and *whuffs*. Such a bizarre sound for animals weighing nine hundred pounds. Then comes a clattering sound—the clashing of antlers as two bulls spar?

Elk cows gather in nursery herds with their yearlings and their calves—dropped in May, now half-grown—feeding on the lush grass of our meadow, drawn also by the salt block placed for cattle under our association's grazing lease.

It is the fall rut, and the bull elk are drawn here by the presence of female elk. Some of the calves and yearlings may be their young from prior years. In this season, the bulls are impelled by the rise of hormones, their necks swollen, their antlers polished and gleaming. We often find the downed remains of low-hanging pine boughs thrashed by them in mock combat as they struggle to release some of the energy and mating angst that drives them. They bugle to lure females, expending great energy to impress the cows with their size and their suitability as sires for the young the females will carry through winter and bear next spring.

Observing the rut through a human lens, people have long described bull elk as gathering harems of cows, but that is a misinterpretation. It is the female who decides whether to accept or reject a bull's advances. And she will not mate until biologically ready. Watching elk gathered for the rut can be almost comical, as a handsome bugling bull lures a cow to him, herding her and her young to keep them with him. But then another bull bugles a challenge and the first turns to respond. And the cow he has worked so hard to attract drifts away, perhaps to mate with a third bull waiting in the wings.

In some hard winters, exhausted by the rut, weak from spending too little time feeding, bull elk do not survive. They literally die for love. But if they have mated with one or more cows, their genes will carry on and they will have succeeded in the ultimate game of life.

We watch in silence this drama of phantoms. Milling shapes, indistinct in the moonlight, filling our meadow. The music of elk families, the bellows of hormone-driven males. Then after some time—who knows what triggers it, some slight creak of the porch floorboards, the faintest of our whispers barely heard but different from the natural music of the night, our scent carried by the slight breeze—they are suddenly all looking at us. Dozens of heads turned our way at the same exact angle as if aligned by a carpenter's square. Then a bull blasts a warning scream, and they are all running west into the timber. An explosion. We hear a frantic jostling of large bodies, faint whimpers of alarm, the calls of mothers to their young. The trampling clatter of solid hooves across fallen trees at the base of Elk Ridge, incredibly loud as stealth is abandoned. The thudding rhythm of those same hooves running up through the pines, up a trail that in the next day's sunlight we will find newly churned by the pounding of many hooves.

They flee—mama cows and antlered bulls, spike bull yearlings and many small calves. In moments the noise has faded up the ridge and the meadow is nearly empty, but not quite. Left behind are two or three young calves, confused, bleating plaintively for their mothers. *The lost calves begin bleating and bawling in high-pitched shrieks and moans, the same sounds we had heard earlier.*

Finally they, too, are gone, whether instinctively following their clan up the ridge or rescued by their mothers, it is too dark for me to know. And the meadow is once again quiet, filled only with moonlight.

———————

CABIN JOURNAL—OCTOBER 2001: *Three-year-old Olivia is becoming quite the pint-sized naturalist. She spots animal tracks readily, knows to follow them along a trail. She knows the paw track of a coyote from the hoof of an elk. Her ears can pick up a far-off bugle, and when we spot the large shapes of elk in the meadow, she does her best to be still and quiet and watch.*

Olivia and I cuddle together on her bed, the comforter pulled around us. She leans against me, holding Fred, her stuffed dog, in her arms. Beyond the cabin walls, the October night is chilly and filled with magic. And I begin a story.

"The elk bugle to call the mountain fairies to play with them in the meadow," I tell her. "When the people have gone inside and the stars come out, the elk glide quietly into our meadow. They are tired of eating grass all day and they miss their friends. And so they call to the fairies, *Come play! Come play!*

"And down from Rattlesnake Ridge and out from among the piñons and ponderosas the fairies come. Their clothes are sewn from the leaves of oakbrush, decorated with the feathery plumes of mountain mahogany and necklaces of piñon nuts. They wear hats made from the bark of the oldest ponderosas, which breaks away in beautiful shapes like jigsaw-puzzle pieces that smell like vanilla.

"The fairies bring special treats for their elk friends—the tender tips and shoots and leaves that grow too high up in the trees for the elk to reach, even with their long necks. But fairies can fly, so reaching the higher branches is no problem for them.

"The elk lower their silky muzzles and nibble the treats the fairies offer. Then they lift the fairies up onto their backs. The fairies climb onto their antlers and swing like monkeys, hang upside down by their knees, slide down their smooth necks like slides at the playground. They ride around the meadow like cowboys—one, two, three fairies lined up on an elk's back. The elk walk slowly with long, elegant strides, because they don't want the fairies to fall, but to a fairy, way above the ground on an elk's back, it feels as if they are galloping! The fairies laugh and whoop; they wave their hats and sing cowboy songs, which they learned by listening to the cowboys back in the days when this was ranchland and cowboys rode their horses all over these hills and canyons.

"Finally the fairies are tired, and they slide down the elk's long legs like a fireman's pole to the ground. All the elk gather around and the fairies tell them stories—elk love hearing stories—of magical beings and the adventures of coyotes and chipmunks and bluebirds, and elk of course, and all the animals in the wild places of Colorado.

"But try as hard as we might, we will never hear their laughing or their stories, because the ears of people cannot hear the voices of the mountain fairies."

Except for Olivia. She hears the fairies. Her eyes are wide and enchanted, seeing sights far from this cozy room. She's out in the meadow beyond the cabin walls, playing in the moonlight with the mountain fairies, swinging from the antlers of elk and laughing with them. And listening to their magical stories.

Elk are, in a way, magical animals. Each spring the males grow an extra pair of limbs. On their heads. These weird limbs are bony, but lack muscle or joints. They don't move or bend, and they fall off when they're no longer needed. Antlers. Male members of the deer family (in Colorado, that's elk, white-tailed and mule deer, and moose) grow antlers each year. Caribou, also known as reindeer, are the only deer species in North America in which females also grow antlers. Except for once a year, on Christmas Eve, we don't have reindeer in Colorado.

In spring, antlers begin to grow from special antler pads on the bull elk's skull. They are soft like cartilage and covered with velvet, a fuzzy living tissue rich with nerves and blood vessels to nurture the growing antler. If velvet is cut, it bleeds just like skin. By late summer, the antlers harden into bone, and the velvet dries and sloughs off. Like a drying scab, the velvet is itchy and the bulls rub and polish their antlers on shrubs and trees. Our land is marked by countless pine saplings with the bark rubbed off on one side and oozing sap, or even snapped in half by over-zealous elk.

Antlers are basically sex organs (OK, not really, but go with me on this), grown for the mating season. So, of course, size matters. A handsome rack on the head of a robust bull tells a cow elk this boy is a prime specimen, a good candidate to father her young. Impressive antlers also intimidate other bulls. The elk version of flexing.

After the rut, antlers have served their purpose. By early spring, they seal off at the base and drop from the antler pad.

This is the time we search for dropped antlers.

At age six, Olivia has moved on from fairies in the meadow. She knows that elk lose their antlers every spring, and she has become a determined antler hunter. Antlers aren't in short supply on our land; we've found many over the years, both elk and mule deer. In antler-shedding season, there is one particular meadow below Rattlesnake Ridge where bull elk seem to discard their antlers. Olivia leads me there.

We're going on an antler hunt . . . We're gonna find a big one . . .

It's a blustery, early April day. We move through brittle grass just beginning to green, our eyes on the ground. An elk antler is three or more feet of curving, hard bone, a main beam with four or more branches on one side, some of those branching further. Every tip, or tine, is called a point, so a six-point bull is a big boy with a fine set of antlers. The direction the tines branch from the main beam tells you whether an antler is a right-side antler or a left.

Olivia is intent on her task, stepping around a patch of prickly pear, searching behind a rabbitbrush. But she is still just six years old, without a lot of height for a vantage point. I realize she is just a few feet away from a pair of antlers lying on the ground together. They are aligned perfectly, side by side, as if the bull violently tossed his head to rid himself of the burdensome headgear and they detached simultaneously, dropping to the ground in the exact placement they were on his skull.

"Let's look over there," I say, leading her toward the antlers in a classic mom subterfuge. Our boots crunch the dry grass as we wander "over there." Then she spies them.

"Mommy, look!" She is so excited, kneeling down beside the antlers. She picks one up, struggling a little because it weighs maybe fifteen pounds. Her eyes glow and she grins ear to ear. "I found two antlers!"

The antler is a six-pointer, nearly as tall as she is, but she manages to carry it all the way back to the cabin. We make quite a parade—proud six-year-old dragging and carrying a six-point antler, eager to show it off to Daddy; me following with the second antler, and our two dogs, Jasper and Rosie, prancing along behind.

Olivia has indeed found a handsome set of antlers. They are a rich brown with whitish tines that are polished and smooth. There are diagonal gashes in the outer parts of the tines. From battling another bull? Reddish-brown blood, still damp, marks the round base of each, as if the bull tossed the antlers just moments before we found them. How free he must have felt with a couple dozen pounds of headgear suddenly gone from his skull! Aaah, no more neck strain.

His relief will be short lived, though, for his body begins growing a new set immediately. What an investment of energy and nutrients an elk makes in antlers—thirty-plus pounds of calcium-rich antlers grown each year, only to be discarded and new ones grown in their place.

We take a photo of Olivia lying on the kitchen floor with the antlers positioned above her head like a bull elk. Later we have a taxidermist mount the antlers on a handsome board, which we display in a prominent spot inside the cabin, a tribute to the monarchs of the mountains that surround us here every fall.

CABIN JOURNAL—MID-JANUARY 1997: *Five inches of snow blanket our meadow, lying even deeper among the trees. So soft and silent and lovely. We follow many tracks of animals—deer and elk, coyotes, rabbits. Tracks of mice—a series of double prints that looks like a zipper: = = = =. I'm surprised at what low branches the deer tracks pass under. They really aren't that big of an animal. Elk are much bigger. We find the low-hanging branches of a ponderosa pine thrashed furiously off the tree and lying on the ground like ravages of war left by a rutting bull elk.*

NEW YEAR'S DAY 2000: *Lots of elk sign—scat, tracks, activity, and lick-marks at salt blocks. One night Rick hears a clacking and mewing—elk cows and calves calling to each other.*

Elk have been part of our cabin story from the beginning. They are on page 1 of our journal, the second animal on our list of mammals (after mule deer) from fall of 1995. Our sightings of them are often at dawn or dusk, or as phantom shapes in the night. On autumn evenings, their eerie bugling comes to us from beyond the trees.

Fall through spring is the season for elk on our land. While people are used to the idea of birds migrating seasonally, they are less familiar with the migration of mammals. In Colorado, elk follow the grass, moving upslope with the spring greening and downslope as winter closes in on the high country. These cloven-hoofed grazers are cousins to cows, and grass is their primary food. They also browse on tender leaves, shoots, flowering plants, even tree bark. Like cattle, elk are ruminants with a multi-pouched stomach. Breaking down cellulose to derive its energy is not an easy process, one the human gut is not good at. After hours of cropping grass, elk must rest and digest, regurgitating their not-fully-processed food to chew and reswallow. The thought of that process makes humans gag, but it works really well for lots of large mammals, including deer, cattle, goats, and sheep. Think of it—an adult bull elk converts grass, *grass*, into nine hundred pounds of animal. Bulk-seeking bodybuilders, eat your hearts out.

The availability of grass, tender browse plants, and nutritious grazing drives the daily and seasonal movements of elk. When spring warms the high country and draws new grass and shoots from the ground, our elk begin their movement west up the Purgatoire River valley into the high country. They summer like well-heeled tourists in the Sangre de Cristo Mountains, in sweeping meadows of thick grass between velvet slopes of fragrant pine.

When autumn begins to close down the high country, browning the grass and burying it beneath snow, they migrate back downslope. They are ready now for the rut, gathering in countless meadows like ours for the age-old rituals.

Except climate change is starting to shake up that pattern.

For the first dozen years we own our land, we see elk regularly from fall through late winter, when the cows, calves, and yearlings move in nursery herds. Groups of ten, fifteen, twenty elk and more gather in meadows and move up and down the ridges.

CABIN JOURNAL—PRESIDENTS' DAY WEEKEND, FEBRUARY 2000: *Mild and dry. No snow on the ground. Elk sign everywhere! Both salt blocks getting use, especially the block to the south.*

About 4:30 PM *we saw a large group of elk out the north window of the cabin. We counted eighteen elk, probably more behind the trees and down the hill, grazing, loafing, and playing in the meadow north of the cabin. Two or three cows, young bulls with small racks and many spike bulls and yearlings. They didn't seem very concerned about us. But when the neighbor who was visiting gently opened the door on the opposite side of the cabin to get his camera from his truck, they spooked. In an instant, they were gone.*

FEBRUARY 18, 2001: *At least eighteen elk cows in Elk Meadow, with more among the trees.*

SEPTEMBER 23, 2002: *Heard a single elk bugle coming from Toro Canyon around 7:40* AM. *Heard another, more distant. Next morning I hear two or more bulls bugling on different sides of the cabin. A large bull elk appeared at the bottom of Elk Meadow. He strides up the meadow toward Elk Ridge and bugles again after disappearing into the trees. A second bull answers from somewhere off to the northeast.*

Wednesday evening I hear a snapping noise to the south. Soon a large bull comes down off Elk Ridge (same bull as this morning?), walks down Elk Meadow along the driveway. Headed for our pond? Behind him in single file come six or seven elk—three cows and the rest calves, one quite small. More bugling heard after dark.

NOVEMBER 2005: *Nearly twenty elk—cows, calves, sub-adults, and spike bulls—moving across the far side of Elk Meadow at midday.*

DECEMBER 26, 2009: *At first light we spot fifteen elk in the far meadow across Toro Canyon—cows and calves. Grazing, moving around in the dawn light, they are like hazy phantoms. By the time the sun is well up they are gone.*

MID-OCTOBER 2010: *We saw a group of ten or twelve elk in the meadow across the road to the east. First I noticed two elk bedded down in the meadow at dawn. Then others joined them to graze a bit later. Eventually the meadow was very busy.*

DECEMBER 26, 2010: *Lots of elk sign. Bedding spots visible in the grass by the old campsite; many scattered droppings—some really fresh—in a meadow higher up on our land.*

Late the next afternoon, we spotted about ten cow elk in Elk Meadow.

DECEMBER 27, 2011: *Snow has settled to a frosty blanket, knee-deep on the frozen pond. Dozens of elk tracks lead across it. The snow in Elk Meadow is churned up where the elk have pawed through the snow to reach the grass.*

Over the years, we seem to see elk less often, and just a few at a time, not the big groups we once saw. Where we once recorded a dozen, fifteen, twenty elk, we now list just a handful. Some years, our journal has no entries of elk sightings at all.

CABIN JOURNAL—FEBRUARY 1, 2013: *Four inches of snowfall—not a lot of elk evidence. Deer and coyote tracks on the road and driveway but only a single set of elk tracks.*

MAY 22–25, 2015: *Five to six cow elk visible in the meadow across Toro Canyon.*

FEBRUARY 2016: *Saw three bull elk while walking on the road atop the high ridge. They crossed the road far ahead of us.*

NOVEMBER 2017: *Elk sign around but we see only five or six animals.*

JANUARY 26, 2019: *Elk tracks here and there, not the tons of elk sign we used to see everywhere.*

This lack of elk could be due to so many factors. Increasing development of natural gas wells and their associated roads in these once-quiet hills that fragment habitat and disrupt movement corridors. More houses, roads, and activity. Better grass and water somewhere else. Even the timing of our visits, which might miss some periods of high elk activity.

But it also seems as if the elk we do see show up later in fall and move out earlier in spring. We hear bugling much less often than we did. By 2020 we have not seen dozens of elk in our meadows—with competing bulls, fickle cows, and mewing calves—in years. I begin to wonder whether the changing climate is playing a role.

Ironically, as climate change is impacting many species negatively, it may be helping elk, at least in the short term. We may be seeing them less often on our land, but not because they're disappearing. Warmer temperatures are lengthening the growing season at higher elevations and reducing the snowpack, allowing elk to stay higher, longer.

But that does not mean this change comes with no downside. A study from the University of Montana and the US Geological Survey found that with less snow in the high mountains, elk are able to more easily find food. They remain in these areas, browsing on shrubs and plants that would typically be buried under snow. That may be good for elk, but nibbling away at the shrubs and trees wrecks habitat for songbirds and other species that depend on it for nesting, food, and cover. Elk seeking winter food strip the bark from aspen, killing the trees; eat alpine willows needed by ptarmigan and songbirds; and trample alpine tundra, killing plants that take centuries to regenerate and increasing erosion.

The study described a "trickle-down ecological effect" that meant fewer birds could use that habitat, and the ones that did were preyed on more heavily because the munched-down trees and shrubs didn't offer them good enough protection. Another nail in the coffin of North American migratory birds.

While less snow might help elk in some ways, warmer temperatures are benefitting parasites and infectious organisms that infest them. Climate change has led to population explosions of winter ticks, hurting elk and other large mammals. Moose in particular are being literally bled to death, single animals sometimes infested by hundreds of ticks, leaving them weak from anemia and vulnerable to all kinds of disease and stresses.

Wildlife disease specialists with Colorado Parks and Wildlife worry climate change may bring a boom of various diseases that affect elk and deer. Many are spread by biting flies and gnats, which benefit from a warmer climate. A CPW pathologist estimated three to four times as many deer died in 2021 from hemorrhagic diseases over previous "normal" years because of drought and delayed freezing temperatures, which would otherwise kill the insects that spread the disease. Drought and delayed winter—a likely *new* normal.

———————

It's been many years since we found shed antlers in that high meadow below Rattlesnake Ridge, though at one time we found them every spring. Olivia is in her twenties now, living in another state, and her days as a crackerjack antler hunter are long gone.

Some Aprils I still search that meadow for antlers. Maybe this year I'll find one.

6

FEATHER CHASE

Under the 3°C warming scenario, changes in climate and vegetation will disrupt ecosystems, including plant and insect communities that provide food, water, and shelter for birds. . . . Birds in Colorado were projected to be particularly vulnerable to two climate threats: extreme heat in the spring and increased risks of wildfires.
—*Survival by Degrees: 389 Bird Species on the Brink,*
  National Audubon Society report, October 2019

CABIN JOURNAL—JULY 2000: *The first night, we hear from the dark across the meadow an airy piping. We cannot identify the sound. The next night, we follow it.*

Cool darkness surrounds us, soft and comforting, as we sit on the deck with a last glass of wine. Overhead the Milky Way lights a trail among stars that are an infinite scattering of diamonds. I trace the dragon tail of Draco in its winding path above Rattlesnake Ridge to the trapezoid of lights that form its head. Much closer above us, a big brown bat chases night insects in a silent, erratic semaphore. From across Elk Meadow a poorwill calls. Then from the dark comes a new sound, a call we do not recognize. It is a muted call, almost a tutting but soft at the edges and steady, a metronome in the night. We cup our ears with our hands to better capture the sound, moving our heads to pinpoint the direction. It seems to come from the narrow meadow beyond the ponderosas that define our old campsite, the grassy space where our arroyo flattens and spreads out before gathering itself again to form the head of Toro Canyon.

The sound is like a series of short whistles but round and airy like the notes of a flute. A bird? Not any of the night birds we usually hear, not the hoot of a great horned, the tremolo of a western screech, the *peent, peent* of a nighthawk. Neither could it be the contact calls of a turkey; those day birds roost now in the upper branches of ponderosas somewhere around us in these rough hills, safe from predators that roam the night.

We retire eventually to bed, but the mystery of this calling voice from the dark stays with me.

The next night we sit again on the deck beneath the stars. From beyond the pines the sound comes again: *whi, whi, whi.* We are ready this time for the chase, and we follow the sound, moving across our meadow with flashlights in hand but unlit. Rick carries two-year-old Olivia, her small face bright with excitement. We move as quickly as we can, stumbling on clumps of blue grama, dodging prickly pear, brushing sagebrush that fills the night with spice. We move around the stand of ponderosas into the meadow where the arroyo widens before funneling into Toro Canyon. The sound is clearer now, louder, an airy piping, rhythmic and steady, and my heart beats a little faster. I think I know now who calls. If only I'm right . . .

We slow our steps, trying to move as quietly as two inelegant, booted humans can across a rough meadow on a moonless night, one of them carrying a toddler. The sound remains constant, seeming to come from a gnarled piñon that stands guard like an elder just where the meadow meets the base of Rattlesnake Ridge as it begins to rise sharply to the north. We pause, not wanting to startle the piper, to make it take wing. We feeble humans may be largely blind in the dark, but this night piper is not and certainly knows we are here. Slowly, setting each foot gently, we step closer. The piping stops. We pause, but there is no rush of wings, no sudden departure of bird from piñon. Though if I'm right, the piper's departure would be as silent as a phantom, only the passage of its shadow against the paler sky to tell us it has gone.

We flick on our flashlights, move the beams slowly to scan the piñon. Perched on a horizontal branch is a wonderful sight—a collection of figures perhaps seven inches tall, each about the size and shape of a small sack of flour. The disc of feathers around each face gives their heads an oversized outline, and above their large eyes a smear of white gleams in the light. Saw-whet owl fledglings!

The young owls sit side by side in a row facing us, siblings posed for a family portrait. I count six shapes, all the same size, like a set of sextuplets on their first day of preschool. They stare back at us with huge unblinking eyes, seeming unafraid and untroubled by us and our lights. Nearby, a lone shape of similar size perches on the very tip of a mullein spear, its streaky breast reflecting dark and light. The piper, I assume. One of the parents. Its calls reminded settlers of the sound made by sharpening, or whetting, a saw. Is the adult owl leading its young ones, newly left the nest, into the night to learn to be saw-whets? The branch is only five to six feet off the ground, very long and level, facing the arroyo and unobstructed by other branches or a brushy understory. A perfect staging spot for a family of owls about to launch out on the hunt.

Olivia turns her face to me, eyes as large and round as the owlets'. I lean my head to hear her excited whisper. "Mommy, bird baby?" I nod and squeeze her hand, and the three of us watch in silence and wonder.

The parent leaves its perch suddenly, the spread of its wings wide, dispersing its few ounces of weight across a large surface area. Low wing loading, like a glider, for quiet flight. The young owls follow, their passage soundless. School is in session.

We follow too, as best we can—awkward, stumbling creatures in such contrast to the elegant owls, so seamless in their union with the night. We find them down the slope, and they are gracious in their tolerance of our intrusive presence. But there is work to be done, lessons to be learned, and the adult bird leads its young out again into the night, the family of owls sifting away across the meadow on silent wings.

The next morning, I am eager to add saw-whet owls to our list of sightings. Bird species number 91 on the list. *Tick.*

I had never thought of myself as a birder. From childhood into college and onward, my focus was mammals. As a zoology undergrad at Colorado State, I worked for a graduate student who was studying coyote behavior with the goal of finding nonlethal deterrents to keep them from preying on sheep. I hand-raised two litters of coyote pups in the Animal Behavior Lab and have an affection and admiration for these beautiful and intelligent animals, even when they are behaving badly.

But most mammals are not easy to see, and as a wildlife writer specializing in wildlife viewing and behavior, it was birds I often wrote about. By the 1990s, birdwatching was the second most popular hobby in America. (Gardening was number one—for those who prefer dirty hands to eye strain.) In the field, I watched, learned, kept notes. I interviewed biologists, learned the life histories of many species, ferreted out the latest research. Still, I was not a birder.

It is birds that share our backyards. Birds that fill our world with color, energy, and song. Birds that many people become obsessed with seeing. And somehow, I found myself there too.

My first book contract in 1992 was for a bird guide—*Watchable Birds of the Rocky Mountains*—but I still didn't call myself a birder. I invested in good binoculars and a spotting scope with a tripod so I could research my next bird guide—*Watchable Birds of the Southwest*—but I still wasn't a birder. Except that I was.

The year the southwest birds book came out we bought our land, and the opportunities to see all kinds of birds in a rich habitat tipped me over the edge. I began keeping an official bird list for the first time, of sightings on our land and the surrounding area.

The feather chase was on.

The list begins.

1995 October—first ten sightings

1. Pygmy nuthatch
2. Black-capped chickadee
3. Mountain chickadee
4. Scrub jay
5. Plain titmouse
6. Sandhill crane (in flight)
7. Pine siskin
8. Western bluebird
9. Common nighthawk
10. Wild turkey

Birding has been compared to detective work, which it is, though the subjects are not guilty of any crime except being interesting, beautiful, and, in some cases, rare. Birders have to be patient and dogged, willing to pursue their quarry up mountain, down canyon, through water, across desert, all while being *discreet*. They have to be able to wait patiently, with little movement or noise. They must be excellent observers, noting telltale details, like eye stripes or "undertail coverts" (yes, that's a thing), which birders call field marks. They must detect movement even in dense foliage, recognize the split-second reveal of a field mark or a characteristic behavior or flight pattern. There is a further price to pay—birders have to cast any fashion sense to the wind and wear silly hats and frumpy field clothes with too many pockets. Plus boring colors like khaki, brown, and olive drab, because most birds see color really well and will *not* forgive fuchsia.

Of course, I had been doing this birdwatching and ID thing for years (except in a cool hat and pants with only the usual number of pockets). Now the cabin offered a convenient base to follow mystery birds through forest and arroyo, or just sit on the deck with binoculars and my morning coffee.

In the first year, we quickly checked off dozens of familiar birds: spotted towhee, lark sparrow, broad-tailed hummingbird, dark-eyed junco, mountain bluebird, common raven, American robin, hairy woodpecker, chipping sparrow, northern flicker, tree swallow, mourning dove, lesser goldfinch, violet-green swallow, red-tailed hawk, Steller's jay.

Brown-headed cowbird, western tanager, white-breasted nuthatch, bushtit, northern oriole, killdeer, common poorwill, rufous hummingbird, great blue heron, turkey vulture, red-winged blackbird, pinyon jay, American coot, Canada goose, common yellowthroat, barn swallow, cliff swallow.

Western kingbird, ash-throated flycatcher, American kestrel, western meadowlark, rock wren, Clark's nutcracker, solitary vireo, pied-billed grebe, yellow-headed blackbird, belted kingfisher, canyon wren, American goldfinch, red-breasted nuthatch, Townsend's solitaire, yellow-rumped warbler, downy woodpecker, sharp-shinned hawk, great horned owl.

The feather chase got a bit obsessive, but it was a fun obsession, finding as many species as we could on our land and the surrounding area. We followed birds through our open woodlands, up canyons, along the creek. We chased songs and sounds, our bird list ticking up.

CABIN JOURNAL—JUNE 1996: *A black shape flies across the meadow beside Long Creek in the characteristic dipping flight pattern of a woodpecker—flap up, glide down, flap up, glide down. It ducks into a nest hole in a snag along the creek, reappears, and flies out to a nearby perch. A handsome woodpecker (aren't all woodpeckers beautiful?). Perched in the sun, it is a blackish green with red face and pale collar. Lewis's woodpecker.* Number 35. *Tick.*

AUGUST 1996: *Chased small gray and white bird all over—it stays high in large piñons and upper parts of ponderosas, repeatedly singing "chu-wee, chu-wert, chu weer," similar to a robin's song but each phrase more defined. Hard to see, moves frequently but not nervously. Finally we get a good enough look to confirm. Solitary vireo.* Number 52. *Tick.*

OCTOBER 1999: *From the deck I catch sight of a milling flock of very large birds in the vast blue airspace between me and Raton Mesa. I estimate twenty-five big white birds, all moving in concert like individual cells forming one giant life-form. The flock flashes in the sun as it banks left, then right, like placards with one light and one dark side being flipped repeatedly—white, obscured, white, obscured. I focus in with the binocs. Blocky body; very big, very broad white wings with black primary feathers. The birds fly with their large heads held upright rather than pointed forward. I recognize the birds, with their short necks and very long bills, famous for their beaks holding more than their belican—American white pelicans.* Number 58. *Tick.*

AUGUST 2000: *Forty lark buntings feeding in the meadow. They must have moved upslope from their shortgrass prairie nesting habitat to feed prior to migration.* Number 67. *Tick.*

NOVEMBER 2000: *Loggerhead shrike in flight and perched. Then it is fluttering around in the open meadow in front of the cabin, hovering like a kestrel. I hope to see this predatory songbird nab a lizard or mouse and skewer it on a thorn or fence barb like an avian Vlad the Impaler, but have to be satisfied with a really clear sighting.* Number 73. *Tick.*

We work our way up Rattlesnake Ridge, through an open woodland of fat Colorado piñons and Rocky Mountain junipers with trunks twisted like abstract sculpture. We're heading for a rock outcrop with an eagle's view across our land. We pass a gnarled juniper with a bare horizontal branch about four feet from the ground, wide and flat on top. Laying motionless on the branch is a bird. A mourning dove, I think. Mourning doves are common neighbors here. Their plaintive coos often greet us, and they were one of the earliest sightings on our list.

We freeze as soon as we see the bird, but the dove takes flight. It all happens so fast I don't get a great look, but something doesn't quite add up. The dove seems a little bigger and fuller bodied than a mourning dove, the plumage paler, the tail too short. But my glimpse is brief and I dismiss it.

On the nearly bare branch, its flat top slightly dished like a saucer, is a sparse collection of twigs. Resting on top are two small white eggs about one inch long. We back away and give the nest a wide berth. I think about going back later for a second look, but we've already bumped the bird off its nest once. Ethics wins out over birding, and I don't revisit the nest site.

Still, the dove nags at me. I have seen mourning doves more times than I could count, and this bird seemed different. My birder radar is up. Hmm.

Over the next few years, we see mourning doves many times, and occasionally one seems a little big and a little pale. Then one Memorial Day weekend, I am coming down the old two-track left over from ranching days that cuts into the drainage below Rattlesnake Ridge. Its flattened roadbed, catching and holding water better than the natural slope, has made a nice seedbed over the decades, and junipers and piñons grow up all along it. I've been out for a ramble, now heading back to the cabin. My mind is on lunch, not birds, but suddenly I startle a bird from a low branch. It's not the same juniper of the mystery dove a few years earlier, but this time I get a better look. The tail is definitely shorter and square, the body heavier than a mourning dove. Most telling, I see flashes of white on the wings as it flies.

Back at the cabin, I consult my various bird guides, but nothing really adds up—the size is wrong or the habitat is off. The closest fit seems to be a white-winged dove, but its range is limited to Texas, New Mexico, and Arizona. This dove feeds on the fruit of saguaro and is important in distributing the seeds. Saguaro, the iconic tree cactus that grows only in the Sonoran Desert of Arizona and northern Mexico. What is a cactus eater doing here?!

Migrating birds blown off course, vagrants turning up in unexpected places beyond their normal range—neither is unusual (and often drive birders into a list-checking frenzy). But I think I've now seen white-winged doves repeatedly amid our piñon-juniper forest at sixty-seven hundred feet, over multiple years and nesting.

The 1992 reference *Colorado Birds*, published by the then Denver Museum of Natural History, lists white-winged doves as "casual in spring summer and fall on the eastern plains," with just nine recorded sightings since 1921. The 1998 *Colorado Breeding Bird Atlas*, which focuses on birds that nest in the state, has no listing for the white-winged dove. But by 2016, the *Second Colorado Breeding Bird Atlas* reports two confirmed incidents of nesting in Colorado, two probable nestings, and seven possible. White-winged doves are increasing in number and expanding their range, the atlas reports, and their nesting in Colorado is part of that northward movement. Other sources report white-winged doves sighted in our area, from Trinidad west and south into northern New Mexico.

Still, not all the life history details add up. White-winged doves are usually found in suburban areas taking advantage of bird feeders or in rural areas around grain fields. Did I mistake my identification? Possibly, but many species defy expectation as they expand their range to exploit resources. Many years our junipers are speckled like a Jackson Pollock painting with blue-green or purple berries. As they range northward with a warming climate, are these southern doves switching from cactus fruit to juniper berries? Their bills are too weak to pry fat-rich piñon nuts from their cones, but maybe they forage for them once the cones open and the seeds fall, like the doves that poke around beneath bird feeders for millet and cracked corn. And there is no shortage of other wild seeds in these hills and meadows.

I rule out other possibilities. The Eurasian collared dove is significantly larger than a mourning dove. I know it very well from my Front Range yard. Rock pigeons and band-tailed pigeons are virtual behemoths compared to the small doves. I've seen band-tailed pigeons on our land a few times, and no way would I confuse one for a dove. At the other end of the size scale, Inca doves and common ground doves—two southern-range doves just beginning to be seen in Colorado—make mourning doves look like behemoths.

A 2009 National Audubon Society study of Christmas Bird Count data found more than half of 305 North American bird species now spend winter farther north than they did forty years ago because of a warming climate. The average was thirty-five miles, but one-fifth of the species moved their range northward one hundred miles or more. Doves are generally nonmigratory, so they are expanding their year-round range northward. The white-winged dove is one of those birds.

―――――――――

CABIN JOURNAL—JULY 1998: *Saw a greater roadrunner down in the grass along the shoulder of Interstate 25 just north of Trinidad.*

I was surprised and delighted when I first saw a roadrunner in Colorado. It turned out a sighting around Trinidad, at the very western edge of the plains at six thousand feet, was not unusual. Southeastern Colorado is at the northern edge of their traditional range, and road-runners are longtime, though uncommon, residents. But a few years later, I saw one beside the county road along Long Creek. Then again a few years after that, in a meadow near our driveway. Neighbors shared their own sightings, usually with a joke about its being followed by Wile E. Coyote.

Climate change is pushing many birds out of our area, but the opposite is happening with roadrunners, as it is with its fellow desert-dweller the white-winged dove. Under the National Audubon Society's climate change projections, with a global temperature increase of three degrees Celsius, roadrunners will expand their range northward by 27 percent by 2080, becoming established throughout southeastern Colorado and north along the Front Range almost to Denver.

CABIN JOURNAL—OCTOBER 2016: *The dogs spook a covey of scaled quail out of the brush near our entrance road. The "scalies" scurry through the grass alongside the road before flying off to the far side of the meadow. I'm guessing there were about fifteen birds.*

Quail of any kind are a delight to watch, with their plump bodies and comical, head-pumping walk. The precocious babies are up on stick legs and following the parents very soon after hatching, a busy cluster of fluffballs milling around mom and dad. Nicknamed "cottontop" for the puff of white feathers of their head crest, scalies are arid country birds, well known in the Southwest and somewhat common residents on the grasslands of southeastern Colorado west to the foothills. We once saw one around a dumpster at a gas station in Pueblo.

Piñon-juniper is not an unheard-of habitat for scaled quail, but below forty-five hundred feet, very rarely up to fifty-five hundred. This sighting? At about sixty-seven hundred feet. And the age of the young means the quail nested in the area.

Under the direst warming scenario, scaled quail are projected to gain range across eastern Colorado and up into Wyoming and to expand into foothills and higher elevations on the Eastern Slope. It seems their range expansion is already happening.

CABIN JOURNAL—MAY 2000: *A mockingbird sings grandly in the meadow by the pond. Perched atop a piñon, it performs for twenty minutes at a time. Incredible repertoire—"chirp chirp chirp, bzzt bzzt, pretty bird, pretty bird, tweetly tweetly tweet . . . " Highly varied song, seems evident it has incorporated the music of different species—pretty warbling phrases, buzzing call of towhees. Just delightful.*

Mockingbirds were the eleventh bird on our list, first seen in June of 1996. I've seen mockingbirds many times on the Eastern Plains, but I was surprised to see them on our land at sixty-seven hundred feet. *Colorado Birds* described them as "very rare in foothills and lower mountains (about five records)" in 1992. But with rising temperatures, mockingbirds are projected by Audubon's climate change model to expand their range northward into Wyoming and upslope into higher elevations.

CABIN JOURNAL—JUNE 2000: *Walking down the driveway we flush a sparrow from the grass. It has white tail edges—lark sparrow. Hidden on the ground beneath a white vetch is a small cup nest of fine grasses. It holds four small white eggs with squiggly black lines. A few nights later, it rains all night, a constant, soaking rain. I worry for the mother lark sparrow on her nest, hoping the vetch gives her good shelter and that the runoff down the meadow will not flood her nest. The next day I check the nest from afar with binoculars. No easy task. She's chosen her nest site well, critical for a ground-nester. Has she abandoned the nest? With much looking, I make out her head—with its handsome bridle-like pattern of black and rufous—behind the fronds of the white vetch that shelters her nest. She is still there, hunkered down, protecting her eggs through whatever comes.*

The lark sparrow was one of our earliest sightings—number 13 on the bird list. In the first ten years, we regularly noted them in the nature journal: *lark sparrows jumping among spreading daisy . . . Several lark sparrows—one singing musically for quite a concert.*

Now I can't recall the last time I saw a lark sparrow. I peeked under that white vetch for several years afterward but never saw evidence of a nest again. The last record of a lark sparrow in the journal is 2005, but we don't record every time we see a bird. Lark sparrow populations appear to be stable in the state, and projections are for them to maintain, even expand, their range, though wildland fire and extreme summer heat are both risks as temperatures increase. Are they still here but we're just missing them?

I begin to wonder what other changes to "our birds" have been going on over the last twenty-five years. Rick and I begin to ask each other, "When did we last see a [fill in the blank]?"

CABIN JOURNAL—MAY 2000: *Pair of western tanagers busy in the big pines down by the old campsite.*

JUNE 2002: *I spot a tanager atop the snag behind the cabin. It is there the next day and the next—it's likely nesting nearby.*

Tanagers have never been common on our land, but a few graced us most years with teasing glimpses of scarlet, black, and sunshine yellow flickering among our bigger pines. Now we see them only occasionally. Tanagers are mountain nesters, but the *Second Colorado Breeding Bird Atlas* found them nesting at higher elevations than before, including one nest at eleven thousand feet. Tanagers are moving upslope, the predicted reaction to climate change for mountain birds. Audubon's most extreme model projects western tanagers would see the loss of up to 40 percent of their range in our area. As our elevation becomes too warm and dry and forests die off, tanagers will likely be limited to higher-elevation forests. No more flickers of tanager red, yellow, and black on our land.

CABIN JOURNAL—MAY 2001: *In Turkey Roost Meadow we hear dueting hermit thrushes in the ponderosas that spill down the north side of our ridge.*

JUNE 2004: *Cool, damp, overcast day in June, barely into the sixties. A hermit thrush sings all day. First from the slope of Rattlesnake Ridge, then from our meadow to the south. Then it flies over the driveway, right over Rick's head, to sing from the trees on the west slope of Elk Ridge behind the cabin. By late afternoon, it is singing again from the wooded ridge. The haunting notes ring like a bell through the cloud-shrouded pines, which by late in the day hang low over the ridges and veil Raton Mesa, misting down among our trees till it feels like the Scottish Highlands.*

The song of the hermit thrush draws me as the song of the Sirens drew Ulysses. I have followed the clear, ringing notes through Colorado's lush, subalpine forests many times, seeking this enchanted singer. I've often gotten close, so close, to this shy thrush only to have its song fade into the distance as the singer flees, leading me ever deeper into the forest. In homage, I use the song of the hermit thrush as my phone's ringtone.

We used to hear hermit thrushes on our land every few years, but our forests are too dry and open to keep them. Under the worst warming scenario, these enticing birds will lose 71 percent of their summer range in the United States and Canada. Under that grim model, the song of the hermit thrush will be lost on our land, though we will likely have them in winter as they move their entire range northward. But thrushes don't sing in winter. And if their breeding range shrinks too much, the song of the hermit thrush will eventually fall silent.

CABIN JOURNAL—JUNE 2007: *We are surprised to have a big flock of pinyon jays in the area. Have not seen them, except for one or two birds, for years. They descend on the meadow, land in the trees behind the cabin, laughing and calling.*

I hadn't seen a group of pinyon jays arrive with all their biker-gang-in-town bravado in many years. We used to hear their mewing, laughing calls ring across the ridges and spot flocks numbering thirty birds, seventy-five birds, even a group we estimated at up to three hundred birds. But in recent years we see only occasional small groups of pinyon jays in flight.

Pinyon jay populations have declined across the species' range, according to Breeding Bird Survey data. Climate change is the likely culprit. Pinyon jays depend on piñon pines, feasting on the seeds, nesting in the branches. Overlay a map of pinyon jay distribution with piñon pine range and they match right up. But piñon forests across the West are dying from drought, heat, wildfire, and the spread of the tree-killing piñon *Ips* beetle. A 2014 study projected a decline of 25 to 30 percent of the pinyon jay's breeding range by 2099. Under Audubon's most severe model, pinyon jay range in southern Colorado east of the Sangre de Cristo Mountains will be almost completely lost.

The pinyon jay–piñon pine partnership goes both ways. These biker-gang jays are basically piñon pine farmers. They gather and cache countless piñon nuts but don't retrieve them all. By essentially planting new generations of trees, they play a vital role in rejuvenating piñon forests. If pinyon jays decline, piñon forests will suffer further.

In April 2022 Defenders of Wildlife, claiming pinyon jay populations may have dropped by as much as 80 percent in the last fifty years, petitioned the US Fish and Wildlife Service to protect pinyon jays under the Endangered Species Act.

CABIN JOURNAL—JUNE 1998: *Watched a pair of mountain bluebirds bringing insects to young in a knothole in a snag near the gnatcatcher nest.*

OCTOBER 1998: *Along the road and surrounding meadows are mountain bluebirds in big groups, sitting on wires or dotting piñons—like scraps of blue sky in the light.*

Mountain bluebirds also seemed to be missing in action on our land after our first few years. We never saw a lot of mountains; they were outnumbered by westerns, and we never confirmed their nesting in the boxes until 2013. Mountain bluebirds are projected to see a 16 percent decline in their summer range in Colorado.

CABIN JOURNAL—DECEMBER 1996: *Not much bird activity. It's chilly but sunny. Only little patches of snow. Most obvious bird is Clark's nutcracker. Many fly here and there, busy foraging in the pines.*

Who couldn't love a bird named "nutcracker"?! Especially in December. We regularly saw them for many years, always just September through December, when they showed up for winter to feed on our ponderosa and piñon nuts. Well named, these quirky mountain jays cache tens of thousands of pine seeds throughout the Mountain West. The seeds they don't find again and eat turn into new trees, so nutcrackers are important for the health of mountain pine forests. But high-elevation forests are facing big trouble with a warming climate, and Audubon's *Survival by Degrees* report lists Clark's nutcracker as one of nine Colorado birds most at risk from climate change. Under the worst warming model, nutcrackers will lose more than three-fourths of their range in North America. Our sightings of nutcrackers seem to be way down, just as our winter temperatures climb.

———————

Cabin Journal—October 2005: *From near where we saw the owlets in the meadow below Rattlesnake Ridge years ago, we hear the whistled-note piping of a saw-whet owl from the darkness.*

My thoughts turn to our saw-whet owls. These tiny, secretive owls are notoriously hard to see. We've heard them only occasionally since the summer we saw the saw-whet family. But the trend is clear: mountain birds, forest-nesting birds, northern-range birds—they're all at increasing risk from a warming, drying, tree-killing climate. Saw-whets? They are currently still considered common and widespread in Colorado. For now.

But the outlook isn't good. Under the worst climate model, saw-whets will lose 72 percent of their current summer range in North America. In Colorado they will stay stable in only the highest mountain forests, possibly even gain range as treelines move upslope and alpine species disappear. But on our land and lower-elevation habitat? They will be gone.

I know I cannot change this trajectory, but I have to do something. In 2021 I build a saw-whet owl nest box for Rick and give it to him for Christmas. In January we install it twelve feet up on the open trunk of a sturdy ponderosa. Will they come? I can only hope.

————————

Our journal is an anecdotal record of sightings, not a controlled, scientific study. We don't write down every bird we see every time. Over the years, as Olivia grows up and family and work demands increase, we don't visit the cabin as often and our stays grow shorter. But imperfect though it may be, I begin to realize our nature journal, the one we've kept for fun, is an unintended record of something happening to our birds, something serious.

But songbirds and doves are not the only species affected by climate change. It is with hummingbirds we notice the most dramatic and visible change.

7

SUMMER'S HUMMERS

BBS [Breeding Bird Survey] detections of Broad-tailed Hummingbirds in Colorado declined an average of 1.9%/yr from 1968 to 2002 and 3.9%/yr from 2002 to 2012, suggesting a long-term and accelerating population decline in the state.
—*The Second Colorado Breeding Bird Atlas*, 2016

Many bird species are predicted to occupy new ranges as a result of climate change. The unassuming Black-chinned Hummingbird is thought to have begun a climate-mediated range shift already.
—*Birds and Climate Change*,
   National Audubon Society, September 2015

CABIN JOURNAL—JUNE 1998: *The broad-tailed hummingbird activity is incredible! Our two feeders have attracted a multitude of manic hummers. We've never seen so many before. Hummer wars go on unabated throughout the day as males spar noisily with one another for control of the feeders. They zip through the air in constant combat, like Star Wars space-fighters, blasting each other with lasers as they hurtle through space.*
   The male broad-tailed hummingbird hangs in the air, his wings moving at such speed they dissolve into a mist around him. The sun sparks his plumage to magnificence; he is a glittering jewel. Suddenly, straight upward he rises, perhaps twenty feet, as if lifted by an invisible cable. Again he hovers. . . . Again he rises. Again he hovers, an endless

moment . . . and then he rises even farther, posing again, a tiny emerald against the blue. And I, far below him on the ground, a mere mortal.

Then the sky releases him and he plummets directly downward, down the invisible column he has climbed, down toward Earth. For a foolish moment, the worry rises in me that he will hit the ground, but of course, there is no need for concern—this dance, this display, is written in his DNA, and he and his kin have performed it over the ages more times than I could ever count.

At what seems like the last moment, but is more like eight feet above the ground, he arcs suddenly upward, carving a *J* in the air. He pauses, hovers . . . and begins to rise, repeating his dance.

If I were in a theater watching a troupe of human dancers, I would applaud, and stand, and applaud some more, for his performance is stunning. But this spectacle is not for me. Somewhere around us perches a female hummingbird, perhaps in the pine right next to me, a tiny bird less spectacular than the dancer, watching with bright eyes. She lacks the magnificent throat gorget he flashes magenta-bright in the sun, yet it is she who decides whether they will mate, whether he measures up as the best father for her young. I imagine her playing coy, pretending not to care, doing a metaphorical hair toss, extending a foot to check her nails. It is for her he dances.

If she accepts him, they will mate, then she will weave a nest of plant down and spider silk no bigger than a shot glass and lay two eggs, each the size of a peanut. She will incubate them for two and a half weeks, and after they hatch—unfurling from their minuscule crypt as impossibly tiny birds with needle bills—she will catch insects for them with her ill-equipped beak, because baby birds need protein to grow. Once they fledge, she will teach them to be hummingbirds, perhaps even lead them to our feeders for an easy meal.

And he? The cad. He will abandon her after mating, off to find another female to impress with his spectacular dance.

I've had a lifelong love affair with hummingbirds, and I first fell in love with them as a little kid at my grandparents' mountain cabin in Estes Park, Colorado. They were the musicians of my childhood summers, jewel-green aeronauts trilling and whirring around me. They defied natural laws, moving at rocket speed, stopping abruptly, hovering in air as no other bird can. My grandmother loved the hummers too, and she took great joy in hanging red-trimmed feeders outside the kitchen and along the deck. Recognizing a kindred spirit in her six-year-old granddaughter, she taught me to make the nectar—"One part sugar to four parts water," she would say, turning her wise face down to my wide-eyed one. "But no red dye, the little birds don't need that." Nana patiently instructed me through the steps, ignoring the sticky pools of sugar water I spilled, then following as I carefully carried the full feeders outside, holding them as I dragged over a chair and climbed up (her steadying hand on my arm) to hang them. In an instant the hummingbirds would arrive, and I was sure they recognized Nana and me and had been waiting for us. She'd put her arm around my shoulders, "Let's let them eat in peace," and we'd back away to watch at a respectful distance.

Hummers are summer birds in Colorado and had migrated for the season when we bought our land in the autumn of 1995. But I knew the mix of ponderosa pines and piñons was good habitat and we would see them when the land awoke in spring. On our first camping trip the next summer, as soon as the tent was up beneath the pines, I set out a feeder.

JUNE 1996: *Hung a hummingbird feeder from the ponderosa near our campsite. Within an hour a female broad-tail visited. She drove away another couple of birds who arrived after her.*

For the two days we camped there, she visited regularly, though "she" might have been another bird. Then it was time to drop camp and head home. Not long after we took down the feeder, the female hummer returned. She flew in and hovered at the spot where the feeder had been—but there was no feeder! She darted left, right, searching that small air space for the disappeared feeder, her posture, attitude, and behavior saying, "It was right here, I know it was!" Off she flew, only to return again, and again, to check. I felt so bad at fooling her, though it was unintentional. I wish I could have left the feeder up, but the risk of rewarding a bear with sugar water was too great. How many more times after we left did she come back to check?

A month later we're back and she's waiting.

JULY 1996: *Female broad-tail checks out the feeder, nervously zipping in for a split-second, then darting away. Finally she calms down and is soon sipping contentedly then perching on nearby twigs to defend her food source.*

Was it the same female? No way to know, but I have fun thinking it is. I love how she has turned into a territorial badass, driving off a male broad-tail from the feeder. Later, a smaller hummingbird, copper colored, comes in for a try, and she drives it away as well. I recognize it as a male rufous hummingbird.

July is the month when this headliner hummer arrives. Rufous hummingbirds are smaller than broad-tails, and in contrast to the broad-tail's emerald green, they are a spectacular copper. Rufous hummingbirds (the term *rufous* comes from Latin for reddish-colored) nest in the Pacific Northwest, Canada, and Alaska, but after their young are out of the nest in midsummer, they begin their southward migration along the Rockies. When rufous hummingbirds migrate through Colorado, they bring more drama and action to our feeders than a gunslinger coming to town.

JULY 1996: *By the next day, dynamics at the feeder have snowballed. A male rufous hummingbird was very tentative around the feeder for the first hour, buzzing around, going perch to perch, flying in to hover without feeding, then flying off.*

He cautiously checks out this new food source, tiptoeing in some-one's else's turf. Finally he seems to settle in for a long drink. I see the water surface bobbing as his tongue darts in and out, rapidly lapping.

What a change. The male rufous has claimed the bar, perching now on various watch posts—here, there, behind. He guards the feeder from all comers. A male broad-tail whistles in to "his" feeder, unaware there's a new boy in town. "Red" immediately flies straight at him and chases him off. When the female broad-tail returns, he chases her way. She keeps trying, though, and when the rufous isn't around, I see her feed for extended visits.

Rufous hummingbirds are birds with attitude. These coppery sprites are pugnacious. Every year they show up, tentatively sample the feed-ers, then claim them as their own. They are the Bad Boys of Summer.

Once the rufous hummingbirds arrive, hanging the feeders triggers a carnival of activity.

JULY 1997: *Clouds of hummingbirds of several species crowd around the feeders. They jockey for position, chattering noisily at each other in what I interpret as hummingbird cussing.*

The air is thick with hovering, whirring birds, the occasional muffled whump as one feathered body bumps another. The ringing whistle of broad-tail wings joins the mechanical whir of rufous wings. They spar and duel, fanning their tails in the hummingbird version of shaking a fist in the other guy's face. A male broad-tail and a male rufous rise straight up thirty or forty feet in an aerial face-off, like jousting heli-copters. When they are so high I can just see a couple of specks in the sky, they pause, suspended. I hear them chittering at each other, talking smack. I imagine the other hummers excitedly calling, *Fight! Fight!* and rushing out of the bar to watch.

Finally one bird flies off and his rival hangs in place watching him retreat—the victor. He returns to the feeder and I learn who won. The rufous.

We think of hummingbirds as cute little sprites, but they are fiercely aggressive and deadly earnest—lions in their own aerial world.

I've never seen that level of aggression and competition around flowers in the wild. It is our fault. By offering calorie-rich sugar water at our feeders, we attract the birds and concentrate them unnaturally. In the wild, hummingbirds are dispersed through an area of multiple food sources. Each flower has just a bit of nectar to offer, and once it's consumed, the bird has to move on to the next flower. But feeders are gigantic flowers that never empty. They don't fade like a wildflower patch, so birds come back to the same location over and over.

In mid to late summer, when the young have fledged and the birds are preparing to migrate, or as northern birds move through, headed south and needing energy, the numbers at our feeders skyrocket. Concentrating the birds leads to Hummer Wars.

June 1998: *The broad-tailed hummers are in constant battle over the feeders. First just two males, but eventually a bunch of birds are wrangling. At least one female stood up to an aggressive male who kept trying to chase her off; she chattered back at him and kept trying to move in to the feeder.*

The girls get into it with each other too. *One female was feeding when a second tried to come in to another feeding port. The first pulled back from the feeder, turned, and fanned her tail, flashing a light and dark pattern. The second bird flew away.*

The tail fan reminds me of a dog's snarl—the curl of the lip, flash of sharp white teeth—a silent but scary threat to "Back off!"

––––––––––

In our second year of feeding hummers, a tiny bird joins the mob, and I recognize the smallest bird in North America—a calliope hummingbird.

Cabin Journal—May 1997: *A male calliope at the feeder! His throat is beautifully streaked with magenta, unlike the solid-color throats of other male hummers. He is noticeably smaller, with a shorter bill; tail looks almost bobbed. Unlike other hummers, his tail doesn't extend beyond his folded wings when he's perched.*

One hummingbird species is smaller than another? Seems a bizarre concept when you're talking about birds the size of a wine cork. But in the universe of tiny birds, a few extra millimeters and fractions of an ounce matter. When more than one species is at our feeders, we can definitely tell the difference.

Like rufous hummingbirds, calliopes do not nest in Colorado but pass through in migration. By far the hummers we see most at the feeders are the familiar, trilling broad-tails I've known since childhood. But we realize a new, unfamiliar hummingbird is also at the party.

AUGUST 1997: *Noticed a different sort of hummer at the feeder. It's bigger than a broad-tail or rufous, with a pale speckled throat like a female. But a small patch of dark purple shows in the middle of its gorget. Instead of trilling when it flies, its wings hum loudly.*

I consult my bird guide. It's an immature black-chinned hummingbird. Male, with just a teasing peek of the shimmering purple throat he will have next spring as a full-grown adult. Black-chins are a common nesting bird in southwest Colorado, uncommon though regular on the east side of the mountains in the southern foothills, including our area. Have they been around but we just weren't paying attention?

By 1999 we are feeding hummers at our newly built cabin. Watching hummingbirds is our TV. Avian sitcoms, with color and activity, noise and pratfalls . . .

CABIN JOURNAL—JUNE 2000: *A broad-tailed hummingbird suddenly hits the front window, probably trying to escape another hummer. But instead of glancing off and flying away, its needle bill is stuck in the screen! I see it splayed against the window, wings and tail spread, trying to pull itself free. Poor bird! I rush outside but before I reach it, the hummer has freed itself and flown off.*

. . . or maybe telenovelas, with love triangles and handsome leading men fighting over love interests. And lots of flexing for the ladies.

JUNE 2000: *Broad-tail males making lots of big, swooping flight displays, carving U shapes in air.*

As if he were on a sugar high, a male hummer suddenly rockets straight up until he's just a hovering speck, then plummets downward at rocket speed in an arc that carries him up the other side. Way up there again, he pauses for an eyeblink, then heads over and whistles earthward again. Then he does it again. And again and again, swinging through the arc like a pendulum. swooping a massive horseshoe in the air over and over.

At the bottom of each arc, he chitters and trills with his wings as he swings by. I think the *J*-display we sometimes see is a version of this. Maybe a male stopping to look around for females or male rivals? I sure can't see any, but they must stand out to his eyes like a beacon.

We see plenty of courtship behavior that doesn't seem so whimsical.

At the bottom of the porch steps, a male cornered a hovering female close to the ground, aggressively zipping back and forth in front her repeatedly, maybe two to three feet off the ground, as if stuck in a holding pattern.

This bird, buzzing back and forth in what's called a shuttle display, is basically flying sideways! He flashes his throat colors and flares his tail—this tail flare isn't a "back off" but an invitation for romance. But she doesn't look as if she's ready to swipe right on Tinder. Her bill points up defensively and follows his movement, left-right, left-right. She seems to be trying to get around him and escape.

His display looks really aggressive, and I feel sorry for the little female as this guy fairly screams, "Go out with me!"

Finally she swipes left, flying off at the first opportunity.

Every year we witness more dramas, and some fun comedies.

In August 2005, we catch seven-year-old Olivia and her best friend, Anna Rose, on video as they hula on the deck dressed in construction-paper grass skirts. But in the video, their hummed music, accompanied by a kitchen spoon ukulele, is drowned out by a cacophony of whistling and buzzing from the hummingbirds swirling all around, zipping between the feeders in the background.

In August 2009, hummers swirl around me in a haze of noise and action as I sit on the deck enjoying an evening glass of wine. My legs are crossed so my bare foot dangles in the air, showcasing my newly polished toenails, which are a lovely magenta.

A female broad-tail flies silently up and taps the nail of my big toe! Not a flower, she realizes. She moves to the next, then the next, all down the row of five toes—tap, tap, tap, tap, tap—then flies off, disappointed that my flower-colored nails are not blossoms.

Because of bears, we bring the feeders inside every evening and take them out early the next morning. There are always hummers waiting for us, or more precisely, waiting for the food we bring. These amazing birds don't hum, hover, and rocket through the air at no cost. The metabolism of a hummingbird burns hot and fast, and after a night of fasting, the birds need food desperately to stoke the roaring furnace. Their hunger makes them less cautious.

CABIN JOURNAL—AUGUST 2007: *I brought the feeders out just after first light and before the sun is above the horizon. Hummers are immediately buzzing around me, attracted by the red of the feeders. I hold out one feeder by the wire hanger, staying as still as possible. Instantly four birds are at the feeder ports, dipping in, pulling back, dipping, pulling back. I can feel the breeze from their wings, flapping so fast they are a blur around each tiny body.*

I don't look at them directly, don't want to frighten them with my staring eyes, so I keep my head slightly turned, watching furtively. They are so close, at the length of my arm. This momentary intimacy with these fairy birds is a gift. The morning light sparks their iridescent plumage to glittering jade, copper, magenta. I can see the tiny barbs of their feathers, one complex layer upon another; their long tongues, like flexible toothpicks, darting in and out of their needle bills as rapid as a sewing-machine needle; their tiny feet curved up against their bellies, sharp claws visible on the end of each minuscule reptilian toe; their bright eyes vigilant against possible threats—me and their competitors at the feeder.

I have held the feeders out for hummers many mornings, brought them close to me like a devotional. But this morning is different. It is late summer, the time of migration, and the tides of season bring an additional gift.

At the four ports of the feeder are four different species of hummingbird at one time—broad-tail, black-chin, calliope, rufous!

Time slows, hovers with the birds. It lasts just a few seconds, all of them posed in a tableau. Then the tension of proximity to other birds, and me, is too much, and they break away. Other birds quickly replace them, but the magic of four species together does not repeat.

———————

Something is happening with the hummingbirds. Our first conscious notation comes in 2008.

CABIN JOURNAL—MAY 2008: *We may have more black-chins than broad-tails.*

AUGUST 2008: *Lots of black-chins, not that many broad-tails. Some rufous and calliopes.*

Were we really seeing fewer broad-tails? Were black-chins growing in number, supplanting broad-tails as our most common hummingbird?

SEPTEMBER 2010: *A fair amount of hummingbird activity—lots of black-chins, some broad-tails. Are we seeing a decline of broad-tails or just broad-tails migrating earlier than they used to?*

So many factors affect what species we notice—time of migration of different species, the amount of rain, the availability of wild foods, how much attention we are paying, the timing and frequency of our visits. We don't note all our sightings every visit. Sometimes we don't manage to record anything.

In the early 2000s, the first climate change models project impacts toward the end of this century, and a lot of us secretly breathe a "not in my lifetime" sigh of relief. But I can't deny the changes we are seeing, though I hold on to hope that we aren't losing our broad-tailed hummers.

Sometimes everything still seems status quo.

JULY 2011: *Tons of hummers—draining feeders, constant humming, whistling, chattering. Four species—broad-tailed, black-chinned, rufous, and quite a few calliope.*

MAY 2012: *Good broad-tail and black-chin hummingbird activity at feeders.*

But eventually the evidence just seems too strong. I even make a note in the journal to research what we seem to be noticing.

LABOR DAY 2013: *Still lots of hummingbirds: black-chins are the most frequent visitors, as they were in July. Some little calliopes. One male broad-tail finally materializes, long after the others. NOTE: find out if the ranges, or numbers, of black-chins and broad-tails are changing, the former expanding and the latter increasing.*

EARLY AUGUST 2014: *Plenty of hummingbird action. All four species but not that many rufous or broad-tail. Black-chins rule these days!*

MID-AUGUST 2015: *Lots of hummer activity. They quickly discover the feeders. I see and hear lots of black-chins. Haven't heard any broad-tail males.*

LATE MAY 2017: *Black-chinned hummers visit our feeders. No sign of broad-tails yet . . . (later) Just heard a broad-tail!*

MID-JUNE 2019: *Mostly black-chin hummers, eventually joined by one broad-tail male.*

JULY 2019: *Modest but steady hummingbird activity at feeders. Saw only black-chinned hummers, no broad-tails.*

EARLY MAY 2020: *Black-chin hummingbirds soon discovered our feeders. . . . Some broad-tail males finally join the black-chins at our feeders two days later.*

Of all the birds that live on our land, the ones we will always see in the warm months are hummingbirds. The reason is no mystery. We draw them to us by feeding them. From April through October, we are pretty much guaranteed to attract hummingbirds if we put out feeders. No chasing them through the forest, following them across the meadow, or counting them in nest boxes. Because we see them year after year, and close up, hummers are the birds we most clearly notice changes in over time. And over the years, we have definitely noticed change.

In the twenty-five years we have been watching, the numbers of all hummingbirds we see, of any species, have gone down.

CABIN JOURNAL—LATE MAY 2021: *Surprisingly low level of hum-mingbird activity at our feeders. Even with four feeders out for four straight days, the activity never picks up beyond a minimal level.*

LATE JUNE 2021: *Modest hummingbird activity at the feeders, mostly black-chinned at first, eventually joined by some broad-tails.*

Most dramatic is the supplanting of the once-abundant broad-tailed hummingbird, the classic trilling hummer of summer in the Rocky Mountains, by the black-chinned hummingbird as our most common and abundant hummer.

The National Audubon Society's in-depth 2019 report on the pro-jected impacts of climate change on bird species, *Survival by Degrees: 389 Bird Species on the Brink*, lists broad-tailed hummingbirds as a bird of highest vulnerability to climate change.

Under the direst climate change scenario, a temperature rise of three degrees Celsius, their model projects broad-tails will see a 69 percent loss of range throughout North America. Even with just a one-and-a-half-degree rise, broad-tails face a 45 percent loss. Under either scenario, broad-tailed hummingbirds will likely disappear from our land, and from 19 percent of Colorado, by 2080, even from some of the highest, coolest elevations in the mountains.

But the black-chinned hummingbird is already expanding its range in our state. *The Second Colorado Breeding Bird Atlas* from 2016 reported it's moving northward and eastward. Reports of black-chins increased 30 percent between the first Colorado bird atlas survey in 1998 and the second. Under the scenario of greatest warming, Audubon's climate model projects black-chins will expand their summer range by 62 per-cent, to across most of Colorado.

And that Bad Boy of Summer, the rufous hummingbird? This copper-colored tough guy may lose 71 percent of its summer range under the worst warming scenario, and 53 percent of its winter range. Which means a lot fewer rufous migrating through Colorado.

Our cabin without broad-tailed hummingbirds. The whistling trill of broad-tails no longer heard in much of Colorado. It's unimaginable.

It breaks my heart.

8

THE SUSTAINER OF LIFE

The megadrought in the American Southwest has become
so severe that it's now the driest two decades in the region
in at least 1,200 years, scientists said Monday, and climate
change is largely responsible.
—"How Bad Is the Western Drought?
  Worst in 12 Centuries, Study Finds,"
  New York Times, February 14, 2022

CABIN JOURNAL—MAY 2002: *Conditions <u>extremely</u> dry! Long Creek is
bone dry in places. Almost no grass greened up, oakbrush not leafed
out. Almost no flowers coming up. Much less bird activity. Fewer insects.
Hummers find the feeders in ten minutes. They're swarming feeders and
jockeying all weekend. We suspect it's because natural food is so reduced.
Four of ten boxes occupied. All bluebirds.*

    *JUNE: Still no flowers, except for a hen and chicks cactus just starting
to bloom in Elk Meadow. That's <u>all</u>! Conditions still <u>very</u> dry. Meadows brown and crackling; not a blade of green grass. Some oakbrush
still hasn't leafed out. At first, we see no butterflies except a lone tiger
swallowtail, checking out a can by the cabin, that seems desperate for
nectar. Finally an orange fritillary fritilles past, but that was it. One
dragonfly sighted.*

    *The elk are unusually scarce on our land, even for this time of year.
Then we see groups of up to forty elk—cows and calves—concentrated down*

along the Purgatoire in the afternoon and evening. Along the river seems to be the only place they can find grass and water.

Our boots send up puffs of dust as we crunch through brittle grass as golden as the coat of a mountain lion. It is high summer, but in the meadows there is more bare dirt than vegetation. In other years the tall leaves of blue grama and western wheatgrass have whished against our pant legs at this season. I would have stopped to pull a feathery seedhead, clamping it in my teeth like some seasoned farmer or tickling four-year-old Olivia's neck with the soft plume until she giggled and brushed it away.

But this year, no tall grasses with handsome seedheads dance in the breeze. No symphonies of wildflowers paint the land. No pink or red or white or purple or sunshine yellow. Only dried grasses and the wizened pads of prickly pear, shrunk to anemic green as the plant withdraws the moisture banked in them. The grasses shelter in the ground within their roots, preserving their moisture. They will need what they have to survive the dry.

Beyond the meadow, the trees present an illusion. The ponderosas and piñons and junipers persist in their rich greens and silvered blues, but the color belies the austerity going on within the plant. The moisture content of the living trees in these forests, a neighbor tells us, is no greater than that of kiln-dried lumber. Clouds of powdery earth billow up when we drive the roads to town, turning trees within one hundred yards of the road dull with a coating of dust.

We've had dry years here before, but we can't remember when the meadows were like this, nothing but dust and crackling grass.

But a turnabout is coming, as if the precipitation gods were just playing a joke.

JULY 2002: *Arriving in late afternoon, we find a very different world from the one we left three weeks ago. I had hoped to see a little green grass, but never imagined we would see so much green in the meadows! Abundant new growth—lush grass in the arroyos and in the meadows sloping down to them. More than a foot tall in spots. Oakbrush is not fully leafed out, but it's finally showing buds. Our meadows are actually <u>green</u>, not brown. It's as if the drought has been forgotten. There is a downside: tons of mosquitoes* *in the evening that torment us. Dinner on the deck becomes "take a bite of food—slap—food—slap—food—stand up, grab plate, run for cabin."*

There is evidence everywhere of recent heavy rains: lots of silty mud has flowed down the draws, often marked by deeply imprinted elk tracks. The pond is still dry, but the willows around it are nicely leafed out.

It's an incredible transformation. By Labor Day the pond is filled for the first time in two years. Four inches of rain accumulate in the gauge over the next three weeks.

As the years advance, we realize this neck-snapping turnabout is the norm. What's known as the summer monsoon usually brings torrents of rain midsummer through fall. In only a handful of years does the monsoon not arrive. As long as that early dry period isn't too bad and the monsoon doesn't come too late, plants and animals can catch up. Unfortunately rain sometimes doesn't come in time.

The West is an arid land. Colorado averages about seventeen inches of precipitation annually. Westerners are used to cycles of wet and dry that can turn on a dime within one growing season. Our shallow little pond, formed by an earthen berm across the Toro Canyon arroyo, is one of countless stock ponds built across the West by ranchers to hold on to water that flashes down dry streambeds with these sudden rains. While the popular image of Colorado is the Rocky Mountains, mantled in snow even in summer, drive south from Denver up and over the Palmer Divide, and you enter a land that begins to have more in common with New Mexico. The Palmer Divide separates the South Platte River drainage from the Arkansas River Valley. Once over it, into Colorado Springs and on south toward our cabin, the vegetation noticeably changes. Cholla cactus, its jointed stems like so many arthritic fingers reaching for the sky, grows in profuse gardens across dry grasslands. Piñon pines appear. You have entered the American Southwest.

Western plants and animals are adapted to this wet-dry roller coaster, where a lush wet year can follow several years of drought or a monsoon relieves an early dry. The most dramatic proof emerges magically from the ground after a good rain.

JUNE 2000: *Everything is bone dry upon our arrival. No pond; virtually no water anywhere except Long Creek. Sunday—the sky swirls purple and*

gray. Then a furious thunderstorm! It dumps rain; it hails. Then it fades. We
walk down the driveway, a stream of water flowing in the channel next to
it and feeding into the wide flat spot behind the small dam built across the
arroyo by some rancher, who knows how long ago. Suddenly, we have a pond!

By evening the pond has blossomed with frogs. Rain draws them from
the ground like spring Johnny Jump Ups. Yesterday the land was still and
dry and silent. Today, a vibrant chorus!

The next evening, drawn by the music of amphibian pied pipers,
we walk to the pond. As we reach the edge, the croaking falls gradually
away until just one frog calls, calls, calls. Then silence.

I imagine all the frogs getting the memo except the last poor schlub who
is still croaking away, dreaming of romance. Until someone hisses, "Sheldon, shut it!" Sheldon looks around—"Whut? Oh!"—and finally shuts it.

We sit quietly, unmoving, casting no shadows across the water, and the
chorus gradually strikes up again. A few voices here, a few more there, then
the entire chorus is at it again, full-throated. The croaking is DEAFENING.
How can these tiny frogs, no bigger than a walnut, render up so much racket?

As the sun goes behind Elk Ridge, we start to see shapes floating
on the water surface. In the dimming light, more and more frogs float
to the surface, legs spread. They seem surreally large. Their lungs are
filled with air, swelling their bodies. They are ready for action.

Arching back, the frog blows air from its lungs into its throat sac,
which balloons enormously with a ridiculously loud croak. Then the frog

inhales the air back into its lungs, the air sac deflates, the head drops down as the body rocks up, and the sides swell out. In and out the air passes, the body arching each time in a mighty croak, forcing air back into the ballooning throat sac. And again and again. By dark, we count at least a dozen active frogs on the surface of the water—rockin' and croakin'—with many more unseen, somewhere in the pond.

Like the springtime songs of birds, the croakings of frogs and toads are love songs, meant to attract females for mating. Frog biologists (technically known as herpetologists) dub this courtship croaking a nuptial chorus, summoning images of doublet-wearing frogs serenading their web-toed, goggle-eyed Juliets while cherubs and bluebirds flutter around. What croaking from yonder window breaks!

Also like birds, different frog and toad species have different songs. Our common chorus frogs are little guys, only an inch to an inch and a half across when full grown. But they have mighty voices. Their ballooning throat sacs act like resonance chambers, magnifying the calls. The song of the chorus frog sounds like someone running their thumb down the teeth of a comb, rising in pitch—*re-eeEE-EET!* Over and over. And when a lotta frogs are seeking love, they make a lotta noise. And noise leads to . . .

JUNE 2000: *After three to four nights of croaking, the frogs suddenly fall silent, though the pond is quite full. Two and a half weeks after the rain filled the pond and started the frogs singing, the pond is full of fat-headed tadpoles.*

This wet-dry roller coaster is a natural cycle, and the plants and animals of the West are adapted to it, but they can only take so much. Water is the sustainer of life. In the Northern Hemisphere, the life cycles of butterflies, bees, frogs, birds, and wildlife have evolved to synchronize breeding with the greening of the grass and the blooming of plants so there is enough food—whether it's nectar, berries, or bugs—to fuel the process and nurture the young.

If the pond stays dry, with no water in which to lay eggs or give a home for tadpoles, the chorus frogs will not emerge. They stay dormant in their sealed hibernacula beneath the pond, waiting, waiting for the rains to come. If the rains don't come, few frogs would survive a second year in hibernation.

If delayed rains cause bluebirds to lay their eggs later, the young don't hatch until the end of August or September and the baby birds won't mature and fatten up enough to survive migration or the onset of winter. Plants will respond to late-season moisture, but it won't help the birds and wildlife that needed it earlier.

Bears are one of those species on the clock when it comes to calories. If nuts, acorns, juniper berries, and grubs arrive too late in the season, bears may not catch up on the fat stores they need to survive hibernation. Hunger will rouse bears that enter hibernation without enough fat. But if they emerge in winter hungry, they are unlikely to find sufficient food and will likely starve anyway.

CABIN JOURNAL—AUGUST 1, 2002: *Midafternoon I saw a bear just across the driveway in Elk Meadow. I was on the porch; it must have known I was there, but it ignored me. The bear seemed sluggish and very scrawny, its dark legs too long for the lean torso bleached blond by the sun. Its coat looked rough and brittle, unhealthy. This is definitely not one of the sleek, glossy-coated bears we see in a good year. Are natural foods scarce because of no rain during the early growing season? The bear rooted around in the meadow—what food is there for this ursine on a deadline to put on pounds and pounds of fat before denning? I'm not sure what food it found, for in a moment, it ambled down the drive and disappeared into the trees.*

———————

Though our land seemed to ride the roller coaster between wet and dry years without crashing, after two decades, we sense a trend toward things growing drier. More summers start out dry and stay that way longer. The monsoon more frequently comes late, or not at all.

2011

JUNE 14: *Lots of stunted dry brown grass. Almost nothing growing in the driveway. The meadows are brown and dusty.*

JULY 31: *We arrive in a big rainstorm. Three inches by Saturday night.*

2012

AUGUST 19: *Only two inches of rain in more than two months.*

2014

MAY 5: *Ground is powdery dry.*

JUNE 18: *Hot, windy, and dry conditions.*

AUGUST 1: *What a wet world! The pond is finally back but no frogs.*

AUGUST 24: *Half inch of rain in three weeks. Pond drying up, covered in green slime.*

2016

JULY 5: *Hot! Dry pond. Grown over with a circle of tall sunflowers in its center surrounded by a ring of foxtail grass.*

2017

JUNE: *The May rains are but a memory.*

2019

JUNE 11: *Dry recently but signs of a wet spring. Tall, lush grass in the meadows.*

AUGUST 30: *Still dry. No monsoon this year.*

The drying trend is not our imagination. A super serious, super detailed, super have-to-read-each-line-slowly-to-be-sure-you-get-it report from the NOAA (National Oceanic and Atmospheric Administration), the *Drought Task Force Report on the 2020–2021 Southwestern U.S. Drought*, lays out the dire and inarguable truth.

"Climatologists predict drought in the Southwest in the 21st century will be driven by drying soils due to warmer global temperatures." Warm temperatures that led to the drought being so intense and widespread are going to continue and probably get worse "until stringent climate mitigation is pursued and regional warming trends are reversed."

The Southwest is just going to keep getting drier, says the report, because of more and more greenhouse gas emissions (mostly from burning fossil fuels). With the addition of greenhouse warming, future droughts will be more intense. Weird to think we'll long for the good old days of run-of-the-mill droughts caused by low rainfall and smaller snowpack, what the report calls "randomly occurring seasons of average to below-average precipitation."

And, of course, if nature suffers, so will people. "Human-caused increases in drought risk will continue to impose enormous costs upon the livelihoods and well-being of the ~60+ million people living in the six states of the US Southwest, as well as the broader communities dependent on the goods and services they produce."

There are many consequences of drought in wild lands. One of them we notice early on.

———————

August 2000. I work my way up the back part of our land, where the meadows grow smaller and the land tilts upward more steeply, climbing the slope of Montenegro. I follow the arroyo up to where it branches, choose the righthand fork, then higher up, choose the righthand again. I move from one meadow to the next, getting gradually higher. Then I am at the base of a rocky cliff we call the Thumb, for its thumbprint shape on the topo map. Other features in Colorado have dramatic names like Lizard Head Peak, Royal Gorge, Mount Massive, the Never Summer Mountains. We couldn't have done better than the Thumb?!

From a distance, the Thumb looks imposing, but up close, its sandstone face is much fractured, marked by ramps of broken and eroded rock. Flat ledges of trapped soil scamper up its face, nurturing determined mountain mahogany shrubs and prickly pear clinging to opportunity. A few baby piñons anchor horizontally in soil built up in rock fractures, their thin trunks bending ninety degrees to grow toward the sky. With careful steps and handholds, the Thumb is an easy scramble.

I scan the cliff, choosing my route. But I notice something new above me along the edge of the cliff—several dead piñons, their silhouettes bony against the sky.

In ten minutes I'm standing atop the cliff edge of the Thumb. It's a wonderful spot, austere and dramatic, projecting above the meadows and pine forests that roll down these hills as far as I can see. The sky is so clear and blue and endless I can't help but breathe it in deeply.

But it's not the view that has my attention. Why have at least eight trees died up here? Some stand forlorn and skeletal, dead long enough to lose all their needles. Others, though obviously dead, are a mix of green and brown. I study the trees, trying to figure out what might have happened. Lightning strikes are common tree killers here, especially in such an exposed location. We've got plenty of lightning-damaged pines on our land, but I see no blackened slashes or scars, no lightning-split trunks.

In places where the bark is missing, the grooved trails left by insects munch-munching into the inner wood meander like calligraphy. As I peer closely, I hear something from beneath the bark. *Chik-chik-chik-chik*. Whatever insects chewed those trails, they're still feasting. That doesn't necessarily mean they are the culprits. Trees are their own micro-ecosystems. They host all manner of invertebrates, which in turn feed all manner of vertebrates. Who are these *chik-chikking*, tunneling munchers, and are they the tree-killers?

Now that my radar is up, I begin to notice more dead piñons. A sturdy, tubby tree outside the cabin's living room window, which has been a staging area for birds coming to our pole-mounted winter bird feeder, gradually turns brown and loses its needles until it stands like a grim skeleton. A much larger piñon in the cabin meadow, its top a favorite vantage perch for bluebirds, also dies. We see dying trees at the edges of the cabin meadow and upper Elk Meadow. In a drainage beyond a small, tree-ringed clearing we call the secret meadow, I find a downed piñon lying on the ground like a fallen soldier, hills of pale sawdust below its horizontal corpse. Big, shiny black ants move with great industry along it, in and out beneath the remaining bark. The soldier ants are about three-quarters of an inch long, the workers less than half an inch. Did the ants move in before the tree died, or after? Were they instrumental in killing it?

At the Las Animas County Fair in Trinidad that fall, I stop at the booth for the Colorado State University county extension office. "Is there something killing off the piñons?" I ask.

The volunteer shares a lot of wonderful information about xeriscape gardening, lists of suitable low-water plants for our area, wildflowers to bring color to your garden all season, noxious weeds to control.

"But the piñons," I say, "We have trees that are dying."

I get only apologetic shrugs. They try to be helpful but have no idea, aren't even aware of a specific problem. I try to find other information but turn up nothing. It is 2000, the dark ages compared to the universe of information available online today, and the story of piñon die-off was just coming to light.

Eventually the growing die-off of piñon pines in Colorado and across the Southwest explodes. It's especially devastating across the Colorado Plateau and Four Corners region. Piñon forests become ghostly stands of skeletal trunks. The loss of critical piñon pine nuts as a wild food threatens many wildlife species. We notice lots of piñons dying along the Front Range south of Pueblo, transforming stands of classic PJ forest—piñon-juniper—to just J.

The culprit, it turns out, is not ants but a tiny pine-bark beetle no bigger than a grain of rice: the *Ips* piñon beetle. Its common name is deceptively charming: engraver beetle, for the elegant, curving galleries it munches in tree bark—as if these rice-sized insects sit long hours at their piñon-tree workbenches, spectacles on their noses, artistically inscribing whorls and curlicues in the wood.

Terrified we will lose much of our forest, we decide in desperation to spray the piñons around the cabin to control the *Ips* beetles. It's a fool's errand, really. We have thousands of trees just on our thirty-seven acres. Do we really think we can do anything by hand-spraying a few trees in our "viewscape"?

I find a plump, healthy-looking fence lizard dead in the meadow, with no apparent injury. Did it eat a poison-laced insect? Now we fear the mushrooming consequences of spraying for pine beetles, killing nontarget animals. We don't spray again.

Like the other species of bark beetles that are killing hundreds of thousands of acres of mountain forest in Colorado—including more ponderosas on our land than we are happy about—the *Ips* beetle has always been around. It's a part of the forest ecosystem, and healthy trees can fight off its worst effects, like humans dealing with common colds. Like a human body scabbing over a cut, pines produce more pitch, or sap, to fill the holes drilled by insect pests.

But trees stressed and weakened by drought can't keep up with the invasion and fight off the pests attacking them. "During periods of below average precipitation and warmer than average temperatures, trees become stressed from a lack of water," reads a 2020 Quick Guide brochure from the Colorado State Forest Service. "Stressed trees have a difficult time defending themselves against beetles and succumb to infestations easier than healthy trees. As more trees become infested, beetle populations increase, resulting in widespread tree mortality." In drought years and when warm temperatures persist, *Ips* clans can go through four generations in a single season. All of them munching and engraving and ultimately killing the piñons.

I learn that as a tree succumbs to an *Ips* infestation, its needles die and drop within a year. As the rest of the tree dies, its wood weakens and structural integrity decays. Within two to three years the trees topple, once-vital Goliaths felled by a mob of tiny Davids. The journal offers clues to when the trees at the edge of the Thumb might have first succumbed to a beetle invasion, two years before I found the evidence.

CABIN JOURNAL—JUNE 1998: *Extremely dry conditions, drought year. Pond is bone dry. Hot days, cold nights. Grass is very sparse, few flowers. Lots of flies, almost no mosquitoes.*

The *Ips* engravers get inactive only once daytime temperatures are consistently below fifty degrees. The journal reveals that consistently cool fall and winter days are more and more rare:

At Christmas 2005, it was sixty-eight degrees at the cabin.

At Halloween 2019, it was eighty-five degrees.

At Thanksgiving 2021, it was sixty-eight degrees.

In late January 2022, sixty-four degrees.

Luckily, we are spared most of the *Ips* beetle scourge, thanks perhaps to our summer monsoon remaining mostly reliable, at least for the moment. In the twenty years since I first found the dead piñons on the edge of the Thumb, we have lost only a few of the thousands of trees on our land.

But an EPA fact sheet from August 2016—*What Climate Change Means for Colorado*—says, "Colorado's climate is changing. Most of the state has warmed one or two degrees (F) in the last century." And from another fact sheet—*Water Availability Throughout the West*—"In the decades to come, rainfall during summer is more likely to decrease than increase in Colorado, and periods without rain are likely to become longer. All of these factors would tend to make droughts more severe in the future."

As I write this in January 2022, NOAA's National Integrated Drought Information System, an interactive website updated weekly, shows our land in western Las Animas County in the D3 Extreme Drought category, which means pasture conditions are terrible, large fires can develop, reservoirs are extremely low, and water temperatures are increasing.

It also means trees will get drier and more stressed.

With an increasingly warm climate, the *Ips* engravers are going to be very, very busy.

———————

"Add the cost of drilling a well," I say to Rick. It's December 1998. The cabin isn't built yet; in fact, we're working on the budget to start building next June.

Of course we would drill a well. You build on rural land in Colorado, a well is your water source. We have "adjudicated" our right to drill a residential well by filing the correct legal papers with the state. We have proven we own the land, drawn a map with the well's approximate location, paid our fee, received our permit. We are ready to rock.

Rick adds $10,000 as a budget line item, and we both groan. Drilling a well will be at *least* $10,000. It can't be helped. Like penstemons and piñons, bluebirds and chorus frogs, we need water to live on our land.

But then the story gets more complicated.

Neighbors in the adjacent canyon, we learn, drilled three times before they hit water. At $10,000 a pop. This isn't Minnesota or Florida, with an enormous water table under everything. You poke your straw down seeking ground water, there is always the risk—a pretty big risk—you'll come up dry. It's why westerners have long used the services of a water witcher to find water before digging or drilling. Think a person holding two arms of a three-armed stick, walking around to "feel" the pull of the water. I was actually OK with that—there are more things in heaven and earth . . .

Instead we decide to trust the local drilling company, a multigeneration family-owned operation that has been at it a long time. We'd gotten to know them slightly—the youngest brother of the family has kids Olivia's age—and we would rely on their expertise.

But then we get more news: "The Austins hit water, but . . . "

Neighbors building a ridgetop house to the west of us began drilling and pretty quickly hit water. Good water, and at not too great a depth. But for insurance, the driller suggested they go a bit farther. Ensure their water supply by dipping "the straw" deeper into the water pocket.

Bad idea. Sending the straw deeper, the driller punched through into one of the many abandoned coal mines burrowed beneath our pine-covered hills. We'd seen a schematic map of the mine once—a vast labyrinth of tunnels and roof-and-pillar chambers, now hollowed of coal.

All that good water in the first pocket they hit drained out into the empty mine.

Then neighbors begin to report the results of their well water quality tests. Not great. High dissolved minerals, dissolved methane outgassing from the coal seams, fracking chemicals used in the drilling of natural gas wells leaching into ground water. And the fracking process that releases the natural gas could also fracture the rock pockets we would hope to tap, and our water would drain away.

The uncertainties about drilling a well lead us to a major decision. We will *not* drill a well. Instead, we install a fifteen-hundred-gallon cistern and have water delivered as the cistern gets low. The first dozen years, that service is supplied by one-man operation Art Trujillo, a local guy who became a valued friend. For years Art charges us just forty-four dollars per one thousand gallons of water delivered. He is limited, he says, by the rules of Social Security. He can't earn more than a certain amount, but he likes to keep working, so he has kept delivering water even at a low rate.

Art is tall, barrel chested. At first he seems a man of few words—when we first meet him, I ask whether he is related to a man I know named Trujillo who worked in the area. "I ain't related to nobody," Art growls. But it is always our habit of chatting with Art while the water cistern fills, and he grows to like us, I guess, and tells us all about the local doings and what his family is up to. Art's voice has the inflection of many Coloradans of Hispanic heritage, and I wonder whether he grew up speaking Spanish, though I don't ask. He speaks in clipped, short sentences, punctuated by "Ya know," as in "The wife, ya know, she don't like it that I'm on the road so much of the day . . . "

We benefit from the Social Security Administration's restrictions on Art's income, but it can't last forever. After Art retires from the water business, we work with other suppliers. The price goes up, but even at its highest—about $150 a delivery—we calculate that in twenty-three years, we've spent less than $7,000 total. And the water is City of Trinidad municipal water. No issues with water quality. And no $10,000 sent down a dry hole (or drained out into the mine).

Relying on delivered water probably wouldn't have worked if we lived at the cabin full-time, though some full-time residents around here haul their own water out of necessity due to poor quality or their wells drying up. But for our small weekend cabin, it has been a good solution. We've always conserved, used our water wisely. And in busy years, when we get down for only a few days every month, we get by on one or two deliveries a year.

We made our decision for purely pragmatic reasons: cost and water quality. Now, climate change is making that decision look even better, not that we celebrate that fact. Rising temperatures and reduced precip mean ground water is not recharging. Yet as the population increases, more wells—"too many straws"—are trying to use a shrinking resource. Wells all over the state are running dry.

It's an affirmation of our choice to not drill a well that I would prefer didn't exist.

9

CAVITY NESTERS

In recent decades, [western bluebird] numbers have declined
over much of the species' range. Provision of birdhouses
probably has not kept pace with loss of natural nest sites.
—*Guide to North American Birds*, Audubon.org

CABIN JOURNAL—MAY 2001: *Three ash-throated flycatchers mill busily about the meadow in front of the cabin. One lands in a tree to
the left of the deck, another atop a mullein, the third on a tubby piñon
behind that bird. They call to each other—pe-PEET! Perched, they show
the characteristic "bumped-up" head of a flycatcher. They fly in to light
on a branch of the piñon holding the cabin nest box. I only see two birds
now. Was the third a jilted suitor? One of the two remaining ash-throats
pops through the entrance hole of the box and stays quite a while. Is it
inspecting the box for a prospective nest? Last year this box held a very successful bluebird nest that fledged three young in June and in mid-July had
a mama bluebird on a second clutch of eggs. Have the flycatchers beaten the
bluebirds to the punch? Later, the two flycatchers fly away from the tree, so
I check the box. Four white eggs with jagged brown streaks, striped end to
end. Ash-throated flycatchers are nesting in the box!*

We started our bluebird trail in 2000, with no thought of any resi-
dent other than bluebirds. The bluebirds needed no convincing. When
we put up our first four boxes, they immediately took advantage of

the opportunity and nested in all of them. A bear took advantage of the opportunity as well, knocking the campsite box from the tree and devouring the inhabitants.

That was our first surprise.

The second surprise came in 2001, when ash-throated flycatchers moved into the cabin box. That year birds nested in five of our boxes—ash-throats in one, bluebirds in the other four.

The next year brought another surprise.

JUNE 12, 2002: *Violet-green swallows actively coming and going at the salt block nest box, popping in and out of the hole! They fly out and drop to the ground repeatedly, and we realize they are gathering grass for nest material. A bluebird flies to the hole of the box, and a swallow attacks it. They flutter down several branches, the swallow still attacking the bluebird, which finally flies off. Interesting because a bluebird, at one ounce, weighs twice as much as a swallow, but I guess it's about attitude. And the imperative to protect your turf. I brush Jasper and leave his fur snagged on nearby prickly pears and rocks for the swallows to gather to line their nest.*

How fun to have swallows nesting in one of the boxes! These aerialists have swooped and hunted around us for years, pitching, wheeling in the air above the meadows, even in the narrow open space between the stands of ponderosas at our old campsite. They are hunting insects, of course, the everyday work of gathering a meal for themselves and their young, but their grace and agility is anything but commonplace.

The swallows maneuver effortlessly through the multiple dimensions of vertical space. Theirs is a godlike ability, really, the gift not just of flight but also of the form and skills to command the air. With long, tapering wings they cut and turn, climb and dive, controlling the dynamics of flight with infinite adjustments of their feathers, their wings, their tails. All done without thought but by instinctual knowledge of the air and the winds and how to be at one with them. Or rather, how to harness those dynamics to their use. How many countless times have earthbound humans paused, as I do, to gaze in wonder at the flight of swallows?

These small swallows are also graced with beauty, for when the light is right, they glow a magnificent emerald-purple, with underparts of purest white. Back in 2000, we noticed violet-green swallows very busy among the trees on the lower slope of Rattlesnake Ridge, and we followed two swallows to where they alighted on a ponderosa, almost certainly near their nest in a tree cavity. My guess is the pair nesting in our nest box are the young, perhaps even the young of the young, of those birds.

JULY 14: *Violet-green swallows are very active at the salt block nest box, flying in and out. We find a bit of white eggshell on the ground beneath the box. Standing close, we hear faint cheeping within. I would love to open the box and check on them but resist the urge.*

Eventually my better self gives way to the curious naturalist, and when I think both adults are out of the box, I take a quick peek. Fully feathered babies! They are noticeably smaller than bluebird young. Violet-greens incubate for two weeks, then brood their young another two to three weeks. These babies are close to fledging.

I'm excited to see them, but the demands of life back home in Castle Rock keep us from visiting again until Labor Day.

SEPTEMBER 2: *Salt block box—three dead nestlings—rather young, lacking flight feathers. These can't be the v-g swallows, they were fully feathered and ready to fly in July. I wonder whether the box was taken over late in the season by bluebirds. Were the parents killed? Or was it too late in the season for such a late brood and did the adults abandon the nest?*

I hold four-year-old Olivia as we peer in the open box. The desiccated birds are not a pretty sight, but she's the child of a naturalist and has seen bones, skulls, and "parts" her whole life. Still, she struggles to make sense of it. "Where is their mommy?"

I wonder the same thing.

The next year, 2003, swallows again nest in the salt block box. Whether it is the same adults or new residents I have no way of knowing. We discover swallows nesting in a second box, one we put up along the driveway facing Elk Meadow. Ash-throats are again in the cabin box, and bluebirds in four other boxes.

That occupancy repeats again in 2004. As well as another surprise.

JUNE 18: *Mixed clutch of eggs! Four white with brown longitudinal streaks (flycatcher), two blue (bluebird). The eggs are about the same size. Twice we startled a flycatcher out of the box. Is it incubating all the eggs? Was this nest occupied by bluebirds and did the flycatchers chase them away, take over the nest, and lay their own eggs in it? Or were these dud eggs that never hatched and the flycatchers are making do around them?*

When we are able to check again, in late July, any evidence has been obliterated by completely different tenants. Pinyon mice have invaded all but two nest boxes, including this one. We do notice a lot of flycatcher activity and calling in the vicinity though.

As more cavity-nesting songbirds discover opportunity in our nest boxes and take advantage of them, things grow curiouser and curiouser. We have learned to identify the occupying species by their nest and their eggs. Bluebird eggs are sky blue, swallows are white, flycatchers are white with brown streaks. Bluebirds build a grassy nest lined with some feathers and animal hair. Flycatchers roll animal hair into balls and build it up with grass and other material. Swallows often use strips of juniper bark, grass, and green lichens in their nests. Recognizing the nests is helpful, but far from perfect and sometimes we just don't know for sure who has been using a box.

Other surprises sometimes greet us.

MAY 2008: *Pond box filled with sticks and twigs, even poking out the hole, but no eggs, young, or even a cup to hold them. Is this a wren's dummy nest?*

We have house and Bewick's wrens on our land, but they have never nested in our boxes. Like Goldilocks, birds choose spaces to nest that are just right, and the entrance holes in these nest boxes are bigger than what wrens need. An over-large hole might allow in predators. But male wrens prevent competition in their territories by building fake nests in any unused nest cavities regardless of entrance size, filling them with twigs and sticks. The messy collection in the pond box is about impressing a mate, not making a place to raise babies.

JUNE 2012: *Tuesday—I opened the pond box to find three large baby birds almost ready to fledge. I rechecked on Thursday. Surprise! No feathery shapes but a single, scaly one, coiled in the back of the box! I levered the side opening higher to let in more light, and a snake head popped up and hissed at me! Yikes! Quickly closed the box and came back with my camera ten minutes later. Slowly opening the box, I see the snake still coiled, but no hissing or threat this time. I'm wondering if it's sluggish from a meal of baby birds. Was able to take several pictures then closed the box. A pair of adult ash-throated flycatchers flutters and hovers, perches nearby for a second, flies toward the box entrance repeatedly, but flutters away again without entering. The agitated parents call nervously, but there is nothing they can do.*

Back at the cabin, I look at the photos. From the pattern of brown shapes on yellow I can tell it's a bullsnake. And though I have sympathy for the flycatchers, whose grown babies have been devoured by a serpent, I can't help but have admiration for the snake. That nest box is nailed to a tree nearly five feet above the ground. How do you climb a tree without arms and legs?! I imagine the bullsnake slowly making its way up the tree, likely entering the box from one of the four open ventilation corners in the bottom of the box, snaking its limbless way smoothly into the nest. Did it surprise an adult bird feeding the babies? Cause it to fly away in panic? Did the adults see the snake making its way up the tree trunk, perhaps fly at it, attacking with bill and wings, trying to drive it away?

Now I've got to research how snakes climb trees. It's not uncommon, apparently. They manage to "grip" the grooves and projections of the tree bark with their scales and belly muscles, similar to a rock climber finding tiny projections on a rock face for fingertip and toe holds. Then they extend part of the body to find more holds, grip those, release the lower spots, and undulate up the trunk.

JUNE 2015: *I see a house wren exit one of the two holes in the decorative birdhouse a friend gave us. It's a fancy, brightly painted two-story box with holes too small for bluebirds, meant more for decorating a patio than hosting bird nests. We propped it in the crab apple tree we planted outside the bedroom window for south-side shade. Never dreamed it would actually attract birds! Throughout the day, I see both wren parents visit the birdhouse, carrying insects to feed their young. They enter and exit via the upper of the two small entrance holes. Nesting material is visible through the lower hole. Twice I saw an adult wren leaving the birdhouse carrying a fecal sac. Sometimes one adult perches and sings from a nearby snag.*

JULY 2019: *Checking the "new" campsite box, on the other side of some ponderosas from the old box, which fell victim to a bear again after seventeen years, we find three tiny hatchlings and three white eggs! Too small to be swallows. What are they? We continue our nest box rounds. As I start to open the cabin box, also replaced recently, a mountain chickadee flies out! It sits in the tree above, calling and agitated. With a quick glance, I count at least four small nestlings with sprouts of wispy feathers. Nest has a grassy base layer but smaller strands than a bluebird, then layer of juniper, then fur strands, but not furballs like a flycatcher, for the top layer. When we leave, the adult flies down and into the box, then out again. I'm sorry, mama!*

Another surprise. Mountain chickadees are common and familiar neighbors here. They were the third species noted on our bird list in 1995. Why are they suddenly using our nest boxes after twenty years?

We could have missed them in past years, of course. We've had plenty of mystery nests, with no eggs or young and no active adults to give better clues to the occupants. But to suddenly have them this summer in two boxes?

Two *new* boxes, I realize. Boxes we purchased online. Are they different from our old standby Colorado Bluebird Project boxes from the Denver Audubon Society? I have my suspicions.

In the fall, when I know the boxes are empty, I go out with a ruler. The old boxes have an entrance hole measuring one and a half inches in diameter. The new boxes? One and three-eighths inches. Mountain chickadees are much smaller birds than bluebirds—their body weight of four-tenths of an ounce makes a bluebird's one ounce seem positively massive. Smaller bird, smaller entrance hole, less chance of predators getting in. Some of our boxes finally meet the chickadees' housing requirements.

Chickadees like a nest hole of one-and-an-eighth- to one-and-a-quarter-inch diameter, say various sources. The entrance of these boxes is quite a bit bigger than that to a chickadee. But the loss of trees to human development has had cavity-nesting bird populations in a tailspin for centuries. Perhaps natural nest cavities are getting even more scarce, so they take what they can.

Mountain chickadees are on the "most at risk" list from climate change, though we haven't noticed any fewer of them. They've always seemed abundant on our land, often tipped comically upside-down, feeding on the drying heads of prairie sunflowers. Under the direst climate change scenario, mountain chickadees face a 16 percent chance of becoming locally extinct by 2080 in their summer nesting range in Colorado. The continent-wide picture is much grimmer—72 percent range loss under the worst climate scenario and only a 2 percent range gain. Which means they won't just move north and be happy. They will eventually die out.

Maybe our nest boxes will do some small part in helping these busy little birds, at least for a while.

———————

Ah-hah! Yikes! Cool! We've opened nest boxes to find unexpected species many times—a different bird, a rodent, nests of wasps, even an I-can-climb-a-tree-without-arms-or-legs reptile. These surprises are the little discoveries that make us smile or gasp or close the box really fast. They keep things interesting.

But there is a surprise that comes only through taking a long view across many seasons. That kind of surprise is harder to see. For us it came when we looked back over twenty years of nest box data. Our records are not rigorous science; they're anecdotal, more detailed and frequent some years, less so in others. But within those handwritten notes may hide a grim warning.

2000: Year one of our bluebird trail—all boxes occupied by western bluebirds. The only year bluebirds occupied 100 percent of boxes.

2001: Bluebirds nested in four of five boxes. Ash-throated flycatcher, a new species, in the fifth.

2006: Three different species in nine boxes—six bluebird nests, one flycatcher nest, two violet-green swallow.

Over the next ten years, bluebirds accounted for an average of two-thirds of the nests in our boxes. Some years as low as only half the nests, others as much as 80 percent.

The last year bluebirds made up at least 50 percent of the nests was 2017.

In 2018, it dropped to 30 percent.

In 2019, the mountain chickadees moved in, and bluebird nests dropped to 17 percent of the total active nests.

In 2020, bluebirds rebounded to 43 percent.

In 2021, mountain chickadees seemed to have moved in, finally, to an old-style nest box, and we had three active chickadee nests, three flycatcher nests, and only one bluebird. That year, the first year of the COVID-19 pandemic, bluebirds made up only 14 percent of the total nests, as if they, too, were affected by the virus.

Nest success moves up and down naturally through multiple years, depending on all kinds of things—drought, rain, insect abundance, competition. But five years of declines for bluebirds is concerning. Since 2017 ash-throated flycatchers have averaged 40 percent of all nests. Is this a trend? Too early to say.

Ash-throats are arid country birds, their population projected to remain stable even under the direst climate change scenario. They are actually projected to increase their summer range by 42 percent.

The 2016 *Second Colorado Breeding Bird Atlas* reported a 4 percent *per year* increase in their population just in our state, with a 30 percent increase in ash-throats since the first Colorado bird atlas in 1998. Ash-throats prefer open habitat to thick forest, so they may be benefitting from climate change as piñon pine forests die off from drought and the spread of *Ips* beetles.

In contrast, western bluebirds are projected to lose 35 percent of their summer range and 64 percent of winter range by 2080 under the direst climate change model. Presently, though, their statewide population is stable or increasing, according to second bird atlas results.

What of those elegant aerialists, the violet-green swallows? They are described as "moderately vulnerable," losing 42 percent of their summer range under the grimmest climate model, 28 percent under the less severe model. Under either scenario, they are likely to disappear from our land.

Between 2007 and 2012, we had only two swallow nests, both in 2009. We have not had a confirmed swallow nest in our boxes since 2014.

Our ostensible reason for a bluebird trail was to help cavity-nesting birds, but really, it was about us. We have loved every minute of it. Peeking in the boxes, anticipating whose nests or bright eggs we might see. Watching the babies mature from smooth, beautiful eggs to pinkies to fledglings pestering their parents for food.

And nearly every year, the dramatic discovery of another box or two bashed down by bears.

The nest boxes allow us a rare privilege—to peer into the intimate lives of bird families. We are like benign landlords, providing nest opportunities but nothing more. Beyond this tiny effort, nothing we do can ultimately control the fate of these species.

As more trees die in western forests from climate change–aggravated drought and insect infestation, cavity-nesting birds will probably benefit. More dead, decaying trees means more holes for nesting and more insects thriving in the trees for birds to feed on. But the benefits will be short term. Eventually the dead trees will fall and rot, or a wildfire will burn through and clear the forest. If and when our land is largely treeless, the bluebirds and all the other cavity nesters will have left us as well.

10

POTSHERDS
AND PIÑON NUTS

There was probably more than one reason the Pueblo people
left the Mesa Verde region in the late AD 1200s. . . . There
was a drought from AD 1276 through 1299. This drought
probably caused food shortages, especially because the population had grown so large. The resulting hardships may have
led to tension and conflict.
—Crow Canyon Archaeological Center website,
 CrowCanyon.org, Cortez, Colorado

It's a Winnie-the-Pooh kind of day, and Olivia and I are two girls
ready for discovery! We head out with butterfly net, bug catcher,
picnic lunch, and enthusiasm. Sunshine fills the meadow beneath a sky
so blue we throw our arms wide and spin till we're dizzy. All the world
is new to a toddler, everything an object of wonder. And though I am
the supposed teacher, Olivia, really, is the one teaching me . . .

. . . that when you catch no butterflies in your net, beetles are just as
cool. That a grasshopper your mommy folds in your hand has scratchy
feet that tickle, and it will fly free when you open your hand. That ant-
hills offer hours of entertainment and when you puff on a goatsbeard,
the seeds drift magically away on silky parachutes. It's all *amazing*!

We watch cloud-animals parade across the sky, rub fragrant sagebrush on our cheeks like perfume. Olivia hides treats in the grass for Jasper the dog to sniff out, never mind that he is right there beside her watching, his nose inches from her hand. He waits politely, then gobbles each treat as soon as she straightens up.

It takes all morning to cross the 150 yards from the cabin to the arroyo. We sit on the bank, five feet above the dry streambed, and happily break out the picnic.

Then I spy something extraordinary among the rocks and cones washed down the arroyo. A fragment of pottery.

Now it's my turn to be overcome with wonder. I jump down into the streambed and pick it up. No boring shard this one (archaeologists call these "potsherds"). It's still vividly colored after who knows how many centuries, with brushstrokes of black and red ochre on pinkish terracotta. Thin-walled and irregularly shaped, the fragment is slightly curved. A piece of a rounded pot. My mind fills with questions. What early Coloradans left this as evidence of their passing? Who made the pot it came from? And when was it left here? Eagerly, I search for more pieces, but this is the only one; in fact, it will be the only one we ever find on our land.

I eagerly show Olivia but she is not impressed. A prehistoric artifact is not nearly as cool as a grasshopper or a floating plant seed. It's not even shiny! Her optimism is transforming into tired toddler tantrum. We head back to the cabin. Olivia is ready for a nap. And I begin thinking about those who have lived on this land before us.

Long before European traders and settlers arrived, the West was a busy place. Plenty of people roamed these hills, hunting, gathering wild plants for food and medicine. There is no persistent water on our land, so people couldn't have lived permanently here. They would have camped along the nearest water source, Long Creek, just a mile away. Was the long-ago bearer of my pottery fragment on her way there? I think she (the bearer was most likely a woman) was making her way with her family or nomadic band over Rattlesnake Ridge, following Toro Canyon down to the creek. Maybe they stopped for lunch, sat on the bank of the arroyo like Olivia and I did. Did she tickle her daughter's cheek with a sagebrush frond, point out a puffy cloud that looked like a cottontail?

They would have had their possessions with them, household goods like cooking and storage pots. I imagine her stumbling as she crossed the arroyo, a bundle slipping from her back, a treasured pot smashing as it hits the earth. "Damn!"—or synonymous early Coloradan curse word— "That was my favorite pot!" She probably paused a moment to grieve this fragment in her hand before tossing it away for me to find centuries later.

My nomadic pot-bearer probably didn't make the pottery she used. There was a robust trading network throughout the Southwest. This fragment resembles Rio Grande glazeware from one of the northern Rio Grande pueblos, dating from about AD 1300 to 1500.

Colorful glazed pottery like this was more than just utility cookware. Did my pot bearer survey the beautiful pottery on display for trade, the way my eyes glisten at the high-end cookware at Williams-Sonoma? Certainly she made her selection carefully, trading piñon nuts or dried meat or some other commodity for it. I'm happy she made the choice of such a handsome pot. Was this a prized serving dish, her version of my grandmother's best china?

This potsherd has lain on our land for perhaps five to seven hundred years, waiting for me to pick it up. It is a connection, a link to an unknown woman, so different from me and yet so alike. Centuries before my European ancestors displaced her descendants from this land, she walked these meadows as I do, rested in the warmth of the sun beneath a stunning Colorado sky, smiled at the view of the far mesa, played in the meadow with her own young daughter.

She enjoyed and made use of what this land offered, as people have in different ways over thousands of years.

I'm standing at Carpios Ridge in Trinidad Lake State Park's picnic area, looking down on the place that a thousand years ago was a scattered community of settled farmers who also hunted and harvested wild foods. They are known as the Sopris Phase people, named after the turn-of-the-twentieth-century coal-mining town Sopris, which occupied the site centuries later. Ironically, both the town of Sopris and the archaeological sites that were once the homes of the Sopris Phase people now lie beneath the waters of Trinidad Lake.

In the 1960s, the Department of Anthropology at Trinidad State Junior College (now Trinidad State College) worked to complete archaeological surveys along the Purgatoire River before the US Army Corps of Engineers dammed the river to control flooding. The dam would create a reservoir—Trinidad Lake—and submerge their thousand-year-old study sites. The clock was ticking.

The archaeologists found abundant evidence of human habitation. Between about AD 950 and 1200, the prehistoric Soprisians built rock houses mortared with clay and plastered over with stucco—very Santa Fe—that could have passed muster today. The houses had multiple specialized rooms—ones for sleeping, for cooking, for doing stuff. The kind of house Ma Ingalls kept hoping Pa would build for her in *Little House on the Prairie*.

The Soprisians grew corn, beans, and squash and harvested wild foods. They didn't make pottery but traded for it with people living farther south in the upper Rio Grande Valley, the ancestors of today's Pueblo people. A lot of potsherds and some intact pots were found at Sopris sites, but they are completely different from our potsherd and were produced much earlier, before the development of colorful, polychrome pottery.

The Soprisians abandoned their riverside homes somewhat abruptly around AD 1200 to 1250. What happened to them? Why did they leave? Were there conflicts, either internally or with outsiders, that meant they could no longer survive in the Purgatoire Valley? Or did a drought push them over a critical survival point?

Tree ring data from other areas of the Southwest show a major, long-term drought between AD 1210 and 1305. The longest, driest period during that time was a sustained drought between AD 1270 and 1295, thought to be a factor in Ancestral Puebloans abandoning the cliff dwellings at Mesa Verde around AD 1300. Did drought also lead the Soprisians to abandon their homes and fields?

Around 1500, a new people moved into the area, people who wove beautiful baskets tight enough to hold water. The Jicarilla Apache became the dominant people in the area. The Spanish called them *Jicarilla*—People of the Basket. For two centuries, they farmed along waterways, hunted, and harvested wild foods. But endless raiding and attacks by Comanches and Utes forced out the Jicarillas by about 1700. The nomadic Mouache Utes became the primary people of the land, moving widely across southern Colorado until being forcibly confined to a reservation in southwestern Colorado in the late 1800s by the US government. Along with the Capote Utes, they are now the Southern Ute Indian Tribe.

So who might my pot-bearer have been? Rio Grande glazeware was made between about 1300 and 1700, so this potsherd wasn't left on our land by a Soprisian. They had already abandoned their settlements by then. Perhaps a Jicarilla Apache or Mouache Ute woman, or some other nomadic person whose passage through this landscape was never recorded.

At the same time the Mouache Utes were roaming the area, an entirely new group of people was moving into the Southwest and learning to live with, and on, the land.

The ghost adobes cluster like phantoms on the far bank of Long Creek, just upstream from its confluence with the Purgatoire River, the relics of a once-vibrant community.

These structures are of much newer vintage than the tumbled rock relics of the Soprisians' homes. This is a Hispanic village called El Rito that dates, based on headstones in the campo santo—the cemetery— from the 1860s.

No one lives here anymore, but this is not a forgotten town. Descendants of El Rito families, even older folks who lived here as children, still live in and around Trinidad. We've seen community members replastering the adobes and restoring the buildings, including a unique building, the morada, built by La Fraternidad Piadosa de Nuestro Padre Jesus, also known as the Penitente Brotherhood. The morada was a meeting house open only to members of the brotherhood.

Hispanic villages like El Rito were small and remote from busier population centers like Santa Fe and Taos. The residents were devout Catholics, but they had no priest or monk to serve their religious needs. The Penitentes formed as a lay brotherhood to try to meet the religious and social needs of isolated villages, though their important role for the community has been overshadowed in the eyes of outsiders by some of their more fervent practices, like flagellation and mock crucifixion. The morada at El Rito is an interesting relic of a unique cultural innovation, but it is just one building. El Rito is a village where early Coloradans lived and worked and raised their families on this landscape for more than a century.

Why did the residents leave El Rito? It wasn't from a changing physical environment. Many El Ritoans had probably worked in the local coal mine, which closed in the 1940s. The building of Trinidad Lake in the late 1970s, which would have isolated El Rito even more, may have spelled the end for the village.

Olivia's hands are wrist-deep in mud. She's five years old and making toy-sized adobe bricks to build a "pretend house."

Sitting in the sun in her little-kid chair, a big bowl full of mud on her lap, she squints at me and grins, showing a missing front tooth, then proudly holds up her muddy hands. Adobe-making is not a mess-free process.

Is she inspired by the adobe buildings of El Rito? Or the time we watched the local community at Ranchos de Taos re-stucco the church with adobe made on-site, hand screening soil and mixing big buckets of adobe mud? Or by the idea of building a dwelling, even a play one, made entirely of local materials harvested from the land?

Be serious. She's inspired by a spunky, independent nine-year-old named Josefina.

We've been reading the American Girl book series—*Josefina, an American Girl, 1824*—about a girl living on a rancho near Santa Fe when it was still part of Mexico. Josefina's hacienda was built of adobe, the classic Southwestern building style of sun-baked bricks made from soil, straw, and water. Now Olivia (with our help) is making adobe bricks to build her own hacienda.

First step: dirt. In the meadow, we dig up a bowlful of soil. True adobe builders screen their soil, like sifting flour, but we settle for picking out any big gravel or sticks. The cool thing is our adobe house is truly "of the land," right down to being the same color. And if we need more materials, we don't have to drive into town to the Ace Hardware. Just dig up more dirt!

Next we cut a bunch of dried grass—blue grama and western wheatgrass—and chop it into small pieces. We mix soil, grass, and water in a big white mixing bowl, experimenting to get the right mixture— not too much water so it's soupy, not too little so it's mealy and won't hold a shape. Olivia dives into this step—at least to the wrists. What kid wouldn't love an excuse to play in the mud, especially when she has her mother fooled that it's educational!

Then comes the work of a true artisan: forming the bricks. Olivia is the craftswoman, taking a glob of adobe, squishing, patting, and shaping it into a brick about two inches long, one inch wide, and half an inch tall. We've found a big sandstone slab flat enough to be the hacienda's base, and Olivia lays her bricks on it to dry in the sun. It doesn't take long beneath this endless blue sky and the southwestern sun. Soon, they're ready for Olivia to build her hacienda.

She lays the first course of bricks in a neat rectangle, leaving a gap for a doorway, then tops it with a layer of mud mortar. The next round of bricks goes on top of the first, but like all good brickmasons, she offsets the bricks so they lie over the seams of the layer below, strengthening the wall. And she interlocks the corners.

The process is really similar to how we laid the log courses for the cabin. We even leave two window openings. When the walls are high enough, I help her span the doorway and windows with flat pieces of wood—lintels. She builds up more adobe courses till we have actual walls and it looks like a real, miniature adobe house!

Now it needs a roof, and our guide is once again traditional southwestern building technique. If this were a real house, we would chisel notches into the top course of bricks and lay a couple of heavy roof beams—vigas. But Olivia is hot, tired, and over her Josefina role play. Plus, it's time for lunch. So we cut to the next step, laying sticks and twigs to span the opening until we have a roof of poles known as latillas. Real adobe roofs would then be thickly plastered with mud, but our roof doesn't need to keep out rain.

Ta da! We have a southwestern adobe "pretend house."

Olivia is off to eat her lunch, sharing her cheese sandwich with Jasper, who is waiting patiently. But I take a minute to marvel at what she's made. Whether a dirt and straw adobe, a sod house on the prairie, buffalo-hide tepee, brush wickiup, or stacked rock house, the people who lived in this area before us built their homes, and made their livings, from what the land provided.

If you know how and when to look, these pine-covered hills are a land of plenty. And the biggest treasure is the nuts of the piñon pine.

"Gonna be a good piñon year!"

It's late summer 1999, and we'd finished building the cabin. Harry was here to pick up the construction dumpster. He's grinning at me from the cab of his truck, waving his arm around at the pines. "Look at those piñons. Loaded with fat cones." *Peen-yones* he pronounces it, different from my *pinn-yuns*.

It's the first big piñon crop since we purchased our land, though we've been too busy with cabin building to pay much attention to pine cones.

Harry's a friendly, chatty guy (a kindred spirit!). His family has lived in the area for generations, he tells me. His people are Hispanos, people of Hispanic heritage who settled in northern New Mexico and southern Colorado before those areas were part of the United States. Gathering piñon nuts is an important cultural tradition for Hispano families. "My family always went out gathering piñons when it was a good year," he says. He's happy to share his tips for the process.

Piñon nuts are full of fat, sugar, and protein, packing enormous caloric punch for their size. All across the Southwest, piñon nuts are a critical food for wildlife and economically valuable to Indigenous and Hispanic residents. And they're delicious. A single piñon tree can produce up to twenty pounds of seeds, borne on the scales of the cone. Ironically, piñon cones are small—maybe two inches across, an inch and a half high—compared to the cones of a ponderosa pine, which are about four inches high and three inches across. But piñon seeds are much larger than ponderosa seeds. A piñon nut is the size of a small peanut (from a tree averaging forty feet in height) while a ponderosa nut is the size of an apple seed, from a pine that can tower up to two hundred feet.

Why would a tree invest so much in producing a seed that big? The payoff comes in seed distribution (in twenty-first-century lingo, supply-chain economics). Need someone or something to disperse your seeds? Offer them a big, fat, juicy nut. In come the birds and squirrels, who carry off the seeds. The raucous pinyon jays that descend on our land like biker gangs are one of the most important distributors (hence their name, obviously). They gather the piñon nuts, eat some, and cache the rest for later. The abundance of the pine nut crop, and diligence of the jays, means there are lots of piñon seeds cached all over the landscape. The uneaten piñon nuts germinate into new, young trees, rejuvenating piñon pine woodlands and helping them regenerate from *Ips* beetle kills.

Pinyon jays have evolved behaviors and physical traits specifically to get the job of collecting piñon nuts done. Their bills are particularly long and sharp to aid in prying into cones to get at the seeds. Unlike other jays, they lack feathers at the base of their beaks to reduce problems from extremely sticky pine pitch (their scientific name, *Gymnorhinus cyanocephalus*, translates to "naked nose bluehead"). In fall, when the pine nuts are ripe, the jays forage busily in large flocks. They have an expandable throat pouch that can hold more than fifty pine nuts. When the foraging group is full up, they call to each other and fly off to their caching area, usually a treeless habitat, where they begin poking the seeds into the ground. They've been observed walking along, side by side, like a group of farm workers sowing seeds, tucking their gathered treasure safely into the ground for leaner times, when food is scarce. Storing stuff isn't helpful if you can't remember where you put it, but pinyon jays have amazing spatial memories. Researchers found they can recall and return to find their cached treasure up to 95 percent of the time. That leftover 5 percent of unclaimed pine nuts is plenty to germinate new piñon pines and keep the cycle going.

If you're wondering why the bird's name is spelled "pinyon" while the preferred spelling of the tree is "piñon," using the Spanish form of an accented *n* pronounced "nya," you'll have to ask the American Ornithologists' Union, which anoints each bird species with its approved common name.

In good pine nut years, which depend on rain, our trees are just like Harry described them—loaded with fat cones. Come mid-September, the cones begin to open, baring the ripe nuts. Now the race is on. Mice, chipmunks, and squirrels gather them. Birds feast on them. And people are busy too. We see trucks pulled off along local roads, families of all ages out gathering piñon nuts. They've got long poles to knock down the cones, drop cloths spread around the base of the trees to catch falling cones, and buckets to carry them home in.

Harvesting piñon nuts sounds easy and romantic. It's not. It's hard, labor intensive, and very sticky work. We find out firsthand a few years later when we go nut gathering.

CABIN JOURNAL—SEPTEMBER 2004: *It's a banner year for piñon nuts! Piñon cones liberally cover the trees. Most trees have cones in various stages of opening up and dropping their nuts. This is the biggest nut crop we've seen in the nine years we've owned our land.*

Art Trujillo arrives to deliver water, and he and Rick have a big piñon nut conversation. Art scoffs at the seven-year piñon nut cycle Rick mentions. "That's nonsense," he says. "Been way more than seven years since the last good crop around here. But it's been the wettest winter in a long time. This year oughta be a good crop."

Art should know. His family homesteaded up Medina Canyon, farther west of here, chipping a hardscrabble existence from the land. They had no car, so once a week his mother would hitch a ride with the rural route mailman down the canyon to the local store to do her grocery shopping. Most of what they ate, though, they produced themselves, including gathering piñon nuts in the fall. "We were poor, all right," says Art, with no trace of self-pity. It didn't hold him back. He's sent all his kids to college. His oldest boy is "a rich computer software engineer on the East Coast," he says, then adds, a trace wistfully, "He don't come home to Trinidad much anymore."

Art is sore from a couple of days of harvesting. He uses the hook-on-a-pole technique to pull down cones. "I don't bother shelling them, just roast the nuts in the shells," he tells us. Set the oven to 350 degrees, then "salt 'em, stir 'em, and leave 'em till they smell roasted." When they're cool, he cracks them with his teeth and eats them from the shell, like sunflower seeds.

Following Art's advice, we spread a big blue tarp beneath the big, fat piñon by the cabin, the one literally dripping with cones, like a Christmas tree loaded with a hundred years' worth of ornaments. We take long mop handles and poles and whack the branches we can reach, triggering a shower of piñon cones.

With about a million cones on our tarp, Olivia and I sit cross-legged in the shade and begin sorting. She is now six years old, and in

her mind she's helping Josefina, the American Girl in 1820s Santa Fe, gather piñon nuts for the family. In my mind we are the pottery bearer and her daughter centuries earlier than Josefina, bent to our piñon nut harvest on a sunny autumn day.

First we gather the loose nuts that have fallen from the cones. Easy peasy. We learn that the chocolate-brown nuts are the good ones, heavy with a fat seed. The pale ones are duds, so light and empty we just flip them away.

Next we corral the cones that are open, with the nuts lying tantalizingly in grooves on the open scales of the cones. We pry out the seeds, trying our best to avoid the *incredibly* sticky pitch. Hmmm.

Now we enter really hazardous territory—opening the closed cones. They are completely covered with pitch. As we pry open the cones, we get our hands and fingers completely glued together. Pitch gets in our hair, on our clothes, and then on everything we touch, like on the door handle and the water faucet when we go in the cabin to clean it off. Yikes! We finally abandon the closed cones to dry and open in the shed, waiting to harvest their nuts for another day.

Following Art's advice, we wash the nuts and spread them in a single layer on cookie sheets (one with sides so the nuts don't escape and roll all over the floor) and roast them in a 350-degree oven for about ten minutes.

We peek in the oven. Are they done? Olivia's eyes are wide. "They smell like cookies!" she says as the kitchen fills with a piney, nutty aroma. Of course, we have to test one or two of them. "Done?" I ask Olivia. She squints her eyes, considering. "Another minute," she says with great authority. Olivia is an experienced baker and a champion taste tester.

When they're the color of a manila envelope, we pull them out and spread them to dry. Soon we are peering into a bowl full of cooled, peanut-sized piñon nuts. Now we must crack the shell to free the sweet, soft nutmeat. We try various tools—pliers, hammer, our teeth—with mixed results. The hammer tap is maybe the best, but it takes Goldilocks-style practice to get it right. That tap was too light and bounced off or sent the nut flying; that one smashed it flat; that one cracked open the shell but left the meat intact. Just right!

When we were excavating our cabin site prior to building, we made an interesting find. A smooth, solid stone that fit nicely in my hand, even seemed to have worn spots where fingers would have gripped it. A mano.

Manos—the word means "hands" in Spanish—are traditional tools used to grind plant foods. I heft this stone in my hand. It's blocky and heavy. Wouldn't want to drop it on my foot. I weigh it at two pounds,

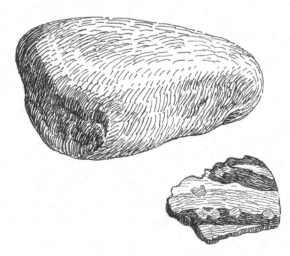

two ounces—plenty solid to crush the shells and grind the soft piñon nutmeats into a moist meal or a paste.

At first, manos were one-handed tools, like this one, which were used to grind seeds, nuts, and wild plants, including medicinal herbs. As southwestern people began cultivating corn and drying it, they needed bigger, heavier tools for grinding, and the two-handed mano rolled across a large stone metate grinding surface came into use. That happened once people lived in settled villages. You couldn't haul around a hundred-pound metate from place to place. Archaeologists, I learn, now use the words *mano* and *metate* for tools used specifically for grinding corn. This tool, then, is better called a handstone.

Nomadic hunter-gatherers used this handstone to grind piñon nuts and acorns, medicinal herbs, pigments. It's a basic technology: harness downward force by applying weighted tool to food product, a.k.a. smashing. No telling how old our stone is. Handstones very much like ours have been dated to nine thousand years ago, and the Utes were using handstones to process food just a few hundred years ago.

Our stone is six inches long and tapers from three and a half inches at the wide end to about one inch at the narrow. It's two inches thick, with a flat, rectangular bottom. Perfect for pounding. Like a prehistoric leatherman, this is a combo unit that can both grind and hammer. What a useful tool. I imagine the owner was just as distressed as the pottery-bearer at loss of her favorite kitchen tool—*No-o-o! Not that one, it works way better than those cheap handstones from Paleo-mart!*

I've felt that way when the handle broke on a favorite chopping knife or the surface of my best sauté pan wore out (twenty-first-century problems . . .). Of course, all I had to do was go to the store or search online and order a new one.

No wonder this stone was that prehistoric cook's fave. She could easily smash piñon nuts with the flat bottom, pick out the shells, then use the broad side to grind the soft nutmeat into a meal or a paste.

Olivia and I decide to give it a try—it's survived maybe one thousand years out here, some twenty-first-century nut smashing isn't going to hurt. We find a nice flat sandstone rock, spread some piñon nuts, and *bash!*, Olivia lays a zinger on the nuts. Half of them fly off the stone into the dirt, but the other half are magnificently smashed into paste on the stone. Success!

I thought we might use our piñon nuts in a fun recipe, but they don't survive the bash-and-snack phase. Nothing like gathering and processing your own wild foods to really learn what it was like to live on the land. Short answer: not easy.

Our piñons are keystone members of the ecosystem here. For wildlife, they offer two critical elements of habitat—food and shelter. And they create an open woodland made use of by countless other species. "It takes a village" is a human spin on the concept of an ecosystem. Here, it is largely the piñons that create the village. But between drought, *Ips* beetles, and wildfire, what will become of this vast piñon "village"?

Harvesting piñon nuts makes me glad that in the twenty-first century, we don't rely on what the land gives us.

Except, of course, we absolutely do.

11

WILDFIRE!

Climate change has been a key factor in increasing the risk and extent of wildfires in the Western United States. . . . Climate change enhances the drying of organic matter in forests . . . and has doubled the number of large fires between 1984 and 2015 in the western United States.
—*Wildfires and Climate Change*, C2ES Center
  for Climate and Energy Solutions website

The [Marshall] fire has forced Coloradans to reassess their thinking about wildfire. About the fire season. About the wildland-urban interface. About climate change. Again. The reminders looked like this: Fire does not need a forest to move. It can travel through the air. The Colorado wildfire season isn't a summer problem. During extended periods of drought, the threat of uncontrollable fire is an every-day-of-the-year affair.
—*The Colorado Sun*, January 2, 2022

CABIN JOURNAL—JUNE 2011: *It's very dry, lots of stunted, brown grass. The meadows are brown and dusty. As we drove down Interstate 25 on the way here, we had views of an impressive plume of smoke rising over Raton Mesa from a wildfire just over the New Mexico line. All through the day at the cabin, smoke rose from just south of Raton Mesa, building into a massive tower that trailed off to the northeast. Luckily the wind carried the smoke away from the cabin, and the sky was otherwise clear blue. By 8:00 PM the wind had subsided and the fresh smoke dwindled to much smaller puffs.*

Wildfire smoke is a herald, sent ahead to announce the looming presence of its ominous master, seen and smelled before the main event. It is a tease, a capricious, shape-shifting visitor that arrives discreetly, preceded by its olfactory calling card.

It is not a welcome visitor, a fact I learned five years earlier in the dry summer of 2006.

I didn't expect any visitors when the dogs and I arrived at the cabin for a much-needed writer's retreat. (I write; they just retreat.) Cabins are great for solitary retreats, and when you're a writer, they're an amazing gift. Work as intently as you need to with no phone, email, or family interruptions. Eat when you're hungry. Sleep when you're tired. Work all night if you're on a roll. Showering is optional, and you can use the bathroom without closing the door—which at the cabin means you have a great view.

June is the perfect month to retreat—endless sky, brilliant sun, soft nights, meadows lush with wildflowers. Except that year, like 2011, the meadows were brown, few wildflowers blooming. *Extremely dry*, I write in the journal, *maybe the driest we've ever seen it.*

Still, I didn't expect a wildfire.

My writing retreat habit is to get up at first light, write until I've run out of words, then take a long afternoon hike. Fire is far from my mind until the second afternoon, when I am making my way up through Turkey Roost Meadow. Like an invisible lure cast out on the breeze, the smell of smoke dances across the landscape until it reaches my unsuspecting nose.

I stop in my tracks. Smoke. Where is it coming from? Turkey Roost is a narrow, grassy clearing that spreads long-ways up the slope behind the cabin. At the upper end I have a view to the southeast. A thin haze of smoke rises from the dark recess of forest down in Long Canyon several miles distant. A phantom drifting among the treetops, so innocent, so silent, coming on little cat feet like Carl Sandburg's fog.

I climb up past the Thumb to the top of Montenegro. The view is spectacular—the soft mounds of the Spanish Peaks to the north, the sharp, snow-capped peaks of the Sangre de Cristos to the west, Raton Mesa to the east. But I'm not here for the view. From this vantage point, the surrounding hills dip and roll all around like ocean swells, deep green waves not of water but of pines and junipers. To the south in Long Canyon, a few miles distant, I spot distinct smoke plumes rising from the forest. Smoke is beginning to fill that part of the canyon, hanging above the trees like low clouds.

This is all private land, not national forest, so I call the Las Animas County Sheriff's Office (whose website is the oh-so-western acronym LASOsheriff.org) to report the fire. Several fires are burning in the area, the dispatcher tells me, and local fire departments are fighting them. I leave my phone number for alerts (to evacuate, though neither of us uses that word), but I'll need to rely on my own judgment for whether and when I should leave. I don't need to pack a go-bag. The small duffle I arrived with, along with my laptop, is my go-bag.

Fighting fires in this landscape is no easy task. Rugged terrain, few roads, arroyos, steep wooded slopes, canyons, mountains. We're technically in the Stonewall Fire Protection District—which our property taxes support and to which we make additional donations because we may *really* need these guys—though the town of Stonewall is at least thirty miles west of here. The Stonewall district is huge, covering

547 square miles of mountains and foothills, forests and meadows. It has an unmanned, satellite fire station in Long Canyon, which basically houses the fire trucks. We know and they know, considering the distance from the little firehouse, the mostly volunteer staff, the need to meet at the unstaffed station then head out on the trucks, that by the time firefighters arrived at our place the cabin would be a pile of ash.

Nervously, I return to the cabin and try to work.

That night, I sit on the deck with a glass of wine, trying to enjoy the soft evening and ignore the smoke plume that's still pluming down in Long Canyon. The familiar night sounds that always make me smile—the singing of crickets, hoots of a great horned, yips of coyotes, calls of a poorwill—can't erase my sense of unease.

I can see a rosy glow rising from Long Canyon. Is the fire growing, or am I just able to see its glow now that it's dark? The wind shifts, and I pray it will not drive the fire across those few miles to us. How fast can a wildfire move? The speed of the wind, I assume.

Jasper wakes me in the night, an unusual thing for him to do. He seems unsettled, and I quickly realize why. In summer we leave all the windows open at night, drawing in the night air, as cool and fresh as water from a mountain stream. But tonight the air has thickened with the oily smell of woodsmoke. I get up and move around the cabin, closing all the windows and turning on the ceiling fan.

The next day a haze of smoke hangs among the trees at the edge of the meadow. I check in with LASO—still fighting fires, but no evacuation orders. The fires are still several miles from here, fairly small and not moving significantly. I look out at the weather vane on the deck, where a male broad-tailed hummingbird perches atop the decorative iron hummingbird that adorns the vane. For him, it's a prime vantage point to watch for interlopers coming in to *his* feeders. The vane is turning lazily, the arrow pointing generally north, meaning the wind is blowing *from* the north. Good. For now, the wind is mostly calm and blowing toward the fires, not from the south where the fires are.

No one in Colorado is a stranger to the smoke haze from wildfires—some within the state, others from as far away as California. In 2002 the Hayman Fire southwest of Denver, at that time the largest and most destructive fire in state history at nearly 138,000 acres, was only about thirty-five crow-fly miles from our home in Castle Rock. It turned the air a thick brown for days, the only consolation being the spectacular sunsets created as particulates scattered the evening light into every color of the rainbow. A questionable trade-off.

By dusk the smoke has worsened. There is an otherworldly thickness to the air. It's hazy and yellowish, as if the air itself were sick. The smell of woodsmoke hangs heavy, reminding me with each breath of the wildfire that looms frighteningly close.

The dogs are restless, pacing around the cabin. They don't like the smell of smoke. It agitates them, makes them nervous. I keep them indoors because of the poor air quality, but a 768-square-foot cabin is a pretty limiting space for two large, active dogs.

I could just pack and go, but the fire is still several miles away in the canyon. I check in repeatedly with LASO, trying to gauge the threat. If I do need to go, I could take a few precautions first, use the fifteen hundred gallons of water in the cistern to wet down the cabin and surrounding grass, maybe buy time for the firefighters. I won't be stupid and overstay, but it's not critical yet, I tell myself. It's not time to evacuate. Yet.

The threat of wildfire is a reality we live with. We've worked to be proactively "firewise" over the years. We chose our building site intentionally, with wildfire in mind. The cabin sits in an open meadow, not among the trees. The meadow extends about two hundred feet below the cabin to the edge of the forest, a nice buffer since wildfires tend to run uphill. The upslope trees behind the cabin are forty to fifty feet away. Every year, Rick mows a firebreak around the cabin thirty to forty feet wide, cutting down dry grass and stalks of mullein and prairie sunflower to reduce fuels.

But there is no denying the mass of fuel all around. These wooded slopes are loaded with deadfall—years and years of trees that have died, even before the *Ips* beetle infestation, plus fallen needles, cones, branches, and a decomposing top layer of soil known as duff. Add to that standing dead timber. Without the occasional small fire over the decades to come through and clean up dead wood, it has built up and up, the unburned forest growing denser and denser. From a wildland fire perspective these are not trees, they are fuel.

Sadly, one of the first culprits in this massive buildup of forest fuels throughout the West is a guy we all grew up loving and trusting. Smokey the Bear.

It shouldn't be news to anybody these days, but wildfire is a natural ecosystem process. Lightning is nature's main fire starter. Indigenous people also routinely set fires—what we would call "prescribed burns"— to manage both forests and grasslands. As European Americans moved into western lands, controlling fires in any kind of coordinated way was not on anyone's radar. But in summer of 1910, the West exploded in more than seventeen hundred wildfires. Driven by strong, dry winds, the fires burned more than 3.1 million acres in the northern Rockies, destroying several towns and killing eighty-five people.

Wildfire could no longer be ignored. It was a terrifying monster, a threat to stamp out every time it raised a flame. Aggressively fighting every wildfire became the overwhelming ethic. The Depression-era Civilian Conservation Corps put thousands of men to work preventing and fighting forest fires. So determined was the US Forest Service to not let fires get out of control that the agency established the "10 AM policy" to suppress every fire by 10:00 AM the next day. Considering the millions of acres of forest in the West and the vast extent of today's wildfires, that was wildly unrealistic.

Then in 1944, the Forest Service and some brilliant advertising folks created the most successful and recognizable government character ever: Smokey Bear. (I prefer to include his middle name—"the"—when referring to him, but his official moniker is Smokey Bear.)

Eighty percent of Americans recognize Smokey. He's way more popular than Uncle Sam, Rosie the Riveter, Woodsy Owl, or Mr. Zip (a Postal Service character who promoted the use of zip codes). With his ranger hat, blue jeans, and fire-fighting shovel, Smokey convinced all of America that "Only YOU Can Prevent Forest Fires!"

We took it to heart. For more than a century, the United States has been aggressively fighting wildfires in the understandable, but short-sighted, view that fire is only destructive, never beneficial. Smokey has been *too* successful. Forests all over the country are much too dense and overgrown, more ready than ever to explode in devastating fire.

As if decades of fire suppression weren't enough, along came climate change to make the party far worse. Rising temperatures and deepening drought further dried trees and soils. People and animals get stressed from harsh conditions, and so do trees. Not seek-a-therapy-group kind of stress, but physical stress. Drought and drying weaken a tree's overall health, strength, and resilience. It leaves them less able to fight off insect pests or withstand disease. After decades of fire suppression, western forests were already growing much too dense, choked with a buildup of deadfall and standing dead timber. Insects, including the piñon-killing *Ips* beetle, always present but in balance in a healthy forest, exploded in population, killing millions of trees and leaving mountainsides covered in brown forests that gradually turned to bare skeletons. A ticking wildfire time bomb.

By the time of the 2002 Hayman Fire, foresters estimated "fuel moisture conditions," (meaning trees, logs, and duff) were the driest in at least thirty years. Deadfall logs along the Colorado Front Range, including our land, had only 5 to 10 percent moisture content. Kiln-dried construction lumber has a 6 to 8 percent moisture content.

But the Hayman Fire was just the beginning. The ten worst wildfires in Colorado history all happened in the twenty-first century. And the three largest? All in 2020.

It's no coincidence. The year 2020 was the second driest ever recorded in state history. And the last thirty years were significantly warmer than any time in the preceding one hundred years. Put another way, the last thirty years were hotter than any time in the lives of my grandparents and great-grandparents.

Anyone who's ever built a campfire knows you can't cut down a living tree and expect the wood to burn. It's too green; it just smokes. But western forests are so dry, living trees do burn. In the Rocky Mountains, subalpine forests—a wet, lush, high-altitude life zone with more than thirty inches of moisture annually—are burning more than at any point in the past two thousand years. Dry trees result from warmer temperatures and low precipitation, two factors that occur regularly in a healthy system. But when the trajectory toward hotter and drier continues over time—decades, centuries, millennia—it indicates a changing climate.

In 2020 I watched, horrified, as the Cameron Peak Fire threatened the forests and mountains and valleys of Larimer County in northern Colorado, where my family lived for decades. Then came the massive East Troublesome Fire, just "over the hill" on the west side of the Continental Divide. Reports came in daily of places I knew well that were burning or under threat.

High winds spread the fire, and suddenly it was burning into the west side of Rocky Mountain National Park. I checked fire status reports constantly. Would Rocky—the park I've loved since I was a little kid, where I found my love of nature, whose official history I wrote in 2015, a park so familiar to me it's like a personal friend—be destroyed?

Then the flames did the unimaginable. They crossed the Continental Divide at twelve thousand feet, high wind carrying embers across miles of treeless tundra. The iconic Bear Lake was threatened, and the town of Estes Park. Friends who live in Estes texted pictures of flames towering in the sky behind the town's storefronts, cars jamming the few roads out of town.

Hundreds of wildland firefighters battling both the East Troublesome and Cameron Peak Fires did heroic work, and the east side of Rocky didn't burn extensively. The park's gateway towns—Estes Park on the east side and Grand Lake on the west—weren't touched, though the fire came literally to the doorstep of Grand Lake, torching three hundred homes on its outskirts. But the threat of wildfire is only getting worse. At 208,663 acres, the Cameron Peak Fire is the largest wildfire in state history. (I hate to add "so far" but it's a painful reality.) It burned from August to December of 2020, also in Larimer County, undaunted by a snow dump of several inches. A fire burning through a snowfall upended the old comfort that May to September is wildfire season and winter snow and cold will protect us. East Troublesome is the second largest wildfire, at 192,560 acres. The two massive fires very nearly merged.

Those two fires destroyed hundreds of vacation and rural homes and nearly burned several towns. The lesson? Unlike Vegas, what happens in wildlands doesn't stay in wildlands. Massive development in the wildland-urban interface (which is exactly what it sounds like) means the destruction potential is huge.

Just before New Year's Eve 2021, the Marshall Fire began in open, drought-stricken grasslands near Boulder, an area that saw just one inch of rain between June and December that year. Driven by winds blowing over one hundred miles per hour, it flashed into suburban neighborhoods. Residents literally had minutes to flee. Though it burned only 6,219 acres—small compared to true wildland fires—the Marshall Fire destroyed 1,084 homes to the tune of close to $2 billion in losses, the most expensive fire in Colorado history and among the top ten most costly in US history. Due to the heroic action of local and county law enforcement officers, who went door to door shouting for people to evacuate, only two lives were lost.

The terrifying thing is, this is likely the new normal.

Our land also lies in an interface, more wildland-exurban than wildland-urban. But like much of the West, we are just one lightning strike or careless campfire away from conflagration. If that happens, we are at the mercy of the wind. Our fate truly depends on which way the wind blows.

So what happened that June of my writer's retreat, when the fires burned so close down in Long Canyon? The wind blew in our favor that year, or rather, failed to blow, which favored us. Firefighters controlled and put out the fires. I didn't have to hose down the cabin. The dogs and I didn't have to flee the flames.

But the next year, the wind showed it wasn't all that trustworthy.

———————

"Mary, come look at this," Rick calls from the deck. He points to the
far end of Rattlesnake Ridge, in the general area of a neighbor's weekend
house. It's the kind of sighting you don't want, a plume of smoke rising
from a rugged, wooded ridge that is basically next door.

We and the neighbor use the same access road, and his gate was
closed and locked when we drove past the day before. Is he there? We
try his phone. No answer. This is serious. We need to check this fire
out—now—so we jump in the car and race toward his place.

The gate is still locked, so we climb through the fence and hustle
up the steep driveway. This graveled roadway is definitely not for the
faint-hearted driver, winding its way up one end of the steep ridge to
his cabin in the sky. About two-thirds of the way up, we round a bend
around a screen of trees . . . and find a car engulfed in flames!

It's like a Fast and Furious movie, except real. A car burning fast
and furiously in these tinder-dry foothills is scary enough, but there's
no one anywhere around. Is someone trapped inside? It's too dangerous to get close, but we don't see any sign of a person. So where is
the driver?

The car sits partway off the gravel drive at a very steep, tight curve—
the kind that offers a rapid trip downhill if you miss it—with the rear
end skewed off the driveway into the brush. It's a white, four-door
hatchback of indeterminate make or model, which I don't even think
to determinate because massive orange-red flames are pouring out of all
the windows. All of them, including the front windshield and the back
window. You know, the windows that don't go up or down. I guess
these went up and never came down.

A dark pillar of smoke billows from the car one hundred feet into
the air, carried to the southwest by the wind. The direction of our cabin.
No trees or shrubs are on fire yet, luckily, but the dry grass beneath and
behind the car is burning. Knee-high flames dance all around the car.

"The gas tank could blow," Rick says, and we back away. There's
little we can do, so we hustle up the driveway, assuming the neighbor
must be at his house if there's a flaming car in his driveway. Before
we get far, we meet him and a woman we don't know coming down.

They're carrying buckets of water, but the reality is a couple of gallons of water are not going to do much.

She was driving up to visit, they tell us, but the car fishtailed on that tight curve and the rear end got stuck in the powdery dirt. Probably afraid she would slide backward down the ridge, she floored it trying to get unstuck. Bad idea in a lot of ways. The vegetation is bone dry. With the rear end of the car stuck in the dry grass, the hot metal of the catalytic converter probably ignited the vegetation, a common cause of grass and brush fires. More than 80 percent of wildfires, actually, are started by people from ignition sources like vehicles, chainsaws, campfires, and cigarette butts.

They've already called the county and the Stonewall Fire District. Trucks from the satellite fire station in Long Canyon are on the way, he tells us. *My god,* I think, *I hope they hurry! Before the nearby piñons ignite. Before the gas tank blows and all fiery hell literally breaks loose across these hills.*

He gives us the code for his gate, and we go back down the drive, open the gate, and wait to flag down the firefighters. We can't see the fire from down at the road, but the column of smoke keeps billowing, like the ominous breath of a crouching dragon. Is it getting bigger? Have any trees caught? Will we suddenly hear a boom as the gas tank explodes?!

The minutes creep past. Where are the fire trucks? We strain to see down the road. Where are they?!

Then we hear the sound of large vehicles, distant at first, ebbing and flowing as they move behind a ridge, then roaring again as they climb a hill toward us. Finally a red fire truck—smaller than a city truck but big enough for the job—comes into view, followed by a rescue truck. What a sweet sight!

The vehicles manage to get up the steep, winding drive—they know how to navigate this terrain. We follow on foot. Amazing the power of a tank of water wielded by experienced firefighters. And the skills and commitment of these folks who turn out to battle fires in this danger-ously dry landscape.

Before long, they have the fire out. All that remains is a scarred hulk of car and a broad, black swath of charred grass. And, I notice, several blackened piñons. I've seen dry pines go up before, and once a fire gets hold and burning well, they explode like a torch. With the southwesterly wind that day, the fire could have flashed in a quick second, spread along and over Rattlesnake Ridge, across Toro Canyon to our cabin. It's a sobering reminder of how vulnerable our little cabin is.

Our journal is peppered with references to smoke and wildfires.

CABIN JOURNAL—JUNE 2000: *Everything is bone dry upon our arrival. No pond; virtually no water anywhere except Long Creek. Extreme fire danger; there have been recent fires west of here.*

JUNE 2002: *Not much wildlife seen, partly because of all the smoke from the huge fire to the west, near Bosque del Oso State Wildlife Area.*

JUNE 2011: *Impressive plume of smoke rising over Raton Mesa from a wildfire just over the New Mexico line.*

JULY 2017: *Lots of smoke, Raton Mesa hazy.*

SEPTEMBER 2017: *Lots of smoke in the air from distant fires all over the west. Stars were barely visible at night through the haze. Fishers Peak is very fuzzy/hazy this morning.*

JUNE 2018: *Ute Park Fire still burning to the south in Cimarron. Hazy skies. Scary.*

The threat of wildfire has always been a part of life in the West. We knew that when we built the cabin in 1999. What we didn't appreciate then—no one did yet except the climatologists who were paying attention—was that the threat of wildfire would grow more serious with each passing year. Between 1980 and 2013, the Colorado governor declared wildfire disasters in Las Animas County six times—1999, 2000, 2006, 2008, 2011, and 2013. The trajectory meshes with the data on increasing drought and rising temperatures. All but one of those years was in the twenty-first century. No wildfire declarations for the first nineteen years, then six in the next fifteen. The wildfire threat to our cabin, and to nearly everywhere in the state and the West, is real and made much worse by climate change.

But big wildfires increasingly come with multiple second acts as huge rains follow, triggering flooding and mudslides. The Grizzly Creek Fire of 2021 in western Colorado not only closed stretches of Interstate 70 through narrow Glenwood Canyon but also led to huge mudslides that have repeatedly buried sections of this critical east-west interstate in twelve feet of mud and rock. A 2022 study projects massive rain events following a wildfire will increase by 50 percent in Colorado by 2100. Uncontrollable wildfires followed by flooding are a rogue genie that's not going back in the bottle.

As the threat of wildfires grows, my thoughts go to the prophetic witches in *Macbeth*. "Something wicked this way comes."

12

IN A FOSSIL FUEL FIELD

We are literally releasing the carbon dioxide that nature had
locked up over a hundred million [years] down below the
Earth. And we're releasing all that carbon dioxide now at a
rate a million times faster [than it accumulated].
—Climate scientist Michael Mann, quoted in
  The Last Hours of Ancient Sunlight, by Thom Hartmann

We pass the stunning artifact on every drive to or from the cabin,
a crumbling ruin lying along the contour of the landscape near
the town of Cokedale. Two mysterious concrete structures, each a dozen
feet high, curve side by side for a quarter mile along Reilly Canyon like
the twin halves of a zipper. They're oddly graceful, receding into the distance as if inviting us to follow and explore their story. With 350 arched
alcoves evenly spaced along their length, they look like nothing so much
as a Roman aqueduct, somehow dropped here in the rough country of
southern Colorado. They resemble ruins from an ancient civilization,
these forlorn structures. Once mighty and important, they lie now forgotten, slowly decaying until one day they will collapse into the earth.

Indeed, these mysterious structures *are* the remnants of a vanished
world, but one much more recent than Ancient Rome. They are coking
ovens, left from coal-mining days, when they turned the black treasure
of coal dug from these hills into an even more valuable commodity,
metallurgical coke used in the making of steel.

The coking ovens shut down in 1947, decades before anyone even heard the terms *fossil fuel* or *climate change*. But the coking ovens were a player in a saga that is far from over. This landscape around the cabin is deeply embedded in the story of climate change, bearing witness to the effects while also being a century-long player in producing the fossil fuels that are the cause.

Cabin Journal—July 2009: *Olivia and I fill our pockets with cubes of coal, crumbled from a coal seam running along the wall of Toro Canyon. They look like dusty black sugar cubes. She takes one and draws stick figures of three people and two dogs on the vertical sandstone rock that lies on top of the coal seam. Instant pictograph.*

It doesn't take a lot of detective work to know this is coal country. Seams of crumbly, black rock are visible everywhere. Small chunks lie on the ground below any exposed cliff or canyon face. In Toro Canyon we stand nose to rock with the coal, easily picking bits of it out of the seam. The coal is pretty compelling, really—fragile and crystalline. It fractures easily into little angular blocks, some with glossy black sides, others a dull charcoal. At times the surface shines iridescent in the light. We draw with our small coal cubes, grind them to make natural pigment paints. I sneak a fist-sized chunk home and slip it in Olivia's stocking the next Christmas, though all I get for my trouble is a tween-age eye roll and a look of long-suffering disdain. And, of course, what else would you do with coal but burn it, so I light a chunk over a candle flame to see firsthand, never having burned coal for fuel or anything else. I'm used to the instant fire from touching a match to campfire tinder, so the coal seems to take forever, but eventually it lights and glows. I realize I have made coal coals.

The pursuit of coal drove the economy around here for decades. Coal mining is an important part of Las Animas County history, with stories of grueling work in labyrinthine mines, labor strikes, coalfield wars, even a massacre of striking miners and their families. But there are also powerful stories of close-knit and vibrant communities, hardworking miners and immigrant families seeking a better life, and, for a few men at the top, vast fortunes made.

For the first half of the twentieth century, coal was definitely king in these parts, and mining left an indelible imprint on the town of Trinidad. The high school mascot is the Mighty Miner, and a dynamic bronze statue of coal miners hard at work stands center stage along Main Street. Nearby is another monument, a massive bronze bird cage, with a bronze bird on a perch within—an homage to the first mine safety worker, the canary in the coal mine.

The coal in these hills is bituminous coal, softer than the anthracite coal mined in Pennsylvania, West Virginia, and eastern coal regions. Once bituminous coal is baked down to remove impurities, the resulting coke burns hot and fast, ideal for smelting iron ore used to make steel. If it seems an odd idea to burn fuel to make fuel, it's the same concept as burning wood down to make the charcoal we grill hamburgers over in the summer.

What was it like to work those coking ovens? I imagine a chilly November day in 1920. It's sunny and bright in southern Colorado, but no autumn sunlight shines through the pall of smoke hanging above Reilly Canyon. Leaning on their shovels, cokers wait by their ovens for the arrival of the next coal train from the three mines owned by the American Smelting and Refining Company (ASARCO). As they wait, they dream of Sunday, the only day of rest in their week, when they can be with their families, attend church, play baseball, and not labor at the coking ovens.

A train loaded with coal chugs into position between the curving lines of ovens. In minutes, the muscular cokers, their clothes grimed with coal dust and tar residue, clamber atop the open cars and begin shovel-ing the coal through openings in the oven roofs. They have 350 ovens to fill, one shovelful at a time, until all fifteen hundred tons of coal arriving that day are moved from train to ovens. Despite the November chill they are soon sweating, pausing a moment to straighten their backs and mop their brows with coal-grimed cloths.

With the ovens full, they seal the tops, close off the Romanesque arches on the sides, and light fires in the stoke holes beneath. For hours the coal will smolder, burning off volatile chemicals and compacting into dense chunks of carbon that burn at up to twenty-eight hundred degrees Fahrenheit, hot enough to melt the iron out of rocky ore.

Coking takes at least eighteen hours, but the cokers keep busy tend-ing the ovens, opening those that are baked and shoveling out the finished coke, reduced now to about eight hundred dense tons. Then they prepare to do it all again.

When their shift finally ends, it's not a far walk home. They live in ASARCO's company town, Cokedale, just one thousand feet from the coking ovens on the other side of what is now State Highway 12.

Cokedale is one of the only surviving coal towns in the state. It was built by ASARCO in 1906 to house the cokers as well as miners working its mines. Usually when mines closed down, the people liv-ing in company-owned towns were evicted and the towns razed. But in 1947, after Cokedale's mines closed, ASARCO offered the town's families the chance to buy their homes for $50 per lot plus $100 per room—$450 for the typical four-room house on a town lot. That equates to $5,700 today. Ready to move forward as a real town, the residents incorporated in 1948. Today Cokedale is a National Historic District with about 150 residents.

We've driven past Cokedale on the way to the cabin a thousand times before curiosity finally propels us off the highway to check it out. We step through a time portal.

Tiny four-room houses line quiet gravel streets. Most are the same vintage, shape, size—square, stuccoed, painted in pale tones, with hipped roofs rising to a central point. Perhaps fifty houses line three straight streets, close together, like a collection of Von Trapp siblings lined up for the Captain's inspection. The homes and the town are trim and well kept. I imagine the children of miners running between the houses, laughing with friends, racing up the steps of the cupola-topped school-house. Mothers hanging out the wash, grandmothers rocking babies on porches, fathers trudging home from the coke ovens or the mines, ready for their dinner.

Now, though, the streets are quiet. I wave to the only two people we see—a woman watering her shrubs and an elderly man keeping watch from a lawn chair in his open garage—and they both nod and wave back.

In its heyday, when the coking ovens raged 24-6, fifteen hundred souls called Cokedale home. They had baseball teams and a dance hall, a church and a schoolhouse. Residents were mostly immigrants from Italy, Greece, and various eastern European countries. An interpretive sign says twenty-three different languages were spoken in this and other coal camps. Work was hard and dangerous, but the vibrant towns built rich communities. Years later, people raised in Cokedale and other mining camps described their childhood as the best years of their lives. Community, extended family, a sense of shared identity—these were the treasured parts of life in a coal company town.

We pass a large, handsome building that is now a museum (you can visit by appointment), and I recognize "the Mercantile," a nice word for the company store. I can't help myself and begin singing in the deepest voice I can muster: "You dig sixteen tons and whaddaya get . . . " Rick rolls his eyes, Olivia slumps in the back seat muttering, "Oh my god, Mom," But Tennessee Ernie Ford's coal miner's ballad has me in its thrall, and I'm not stopping. " . . . another day older and deeper in debt. St. Peter don't call me 'cause I can't *gooooooooooo* . . . " I hold the last note, looking expectantly at Rick, hoping he relents before I run out of breath. He's used to humoring his cornball wife, so finally he joins me in the thundering climax, " . . . I OWE MY SOUL TO THE COMPANY STORE!!!"

If we had wanted to stop for a soda or use the facilities, we would have been out of luck. There are no stores here at all—no 7-Eleven, no Starbucks, no Safeway. It's kind of refreshing. This is a place where people *live*, and gentrification has not found its way off the highway to the little town still hovering in another era.

At the edge of Cokedale, a prehistoric behemoth lies buried with only its back visible above ground. At least that's what it looks like to me. Right across the highway from the coking ovens, a massive black mound nearly a half mile long towers above the road. Long and narrow, with smooth sides and a central spine, it is another relic of coal-mining days. This black hill on which little grows is a slag heap, a massive mound of waste from the coking ovens. On chilly days we sometimes see steam rising from the slag, the breath of a prehistoric behemoth that is only sleeping and might one day awaken and rise from the ground.

Or perhaps the behemoth has already awakened and now hovers in the atmosphere as a greenhouse gas, bringing changes even greater than when it was first mined from the Earth in the previous one hundred years.

"The smoke and gas from some ovens destroy all vegetation around the small mining communities." These words were written in 1911 by W. J. Lauck of the US Immigration Commission, not about the Cokedale coking ovens but similar ovens in the coal country of Pennsylvania.

Another observer, Charles Van Hise, the president of the University of Wisconsin, reported in his book *The Conservation of Natural Resources in the United States* "long rows of these beehive ovens from which flame is bursting and dense clouds of smoke issuing, making the sky dark. By night the scene is rendered indescribably vivid by these numerous burning pits. The beehive ovens make the entire region of coke manufacture one of dulled sky, cheerless and unhealthful." Then he added, "There is no possible excuse for the continuance of the use of the beehive oven. Because capital is invested in beehive ovens is no adquate reason for disregarding responsibility to the future, even if there were a financial loss." This was in 1910.

Now I have a better understanding of the shocking photos I've seen of this area in the early twentieth century—mountain slopes denuded of trees, arroyos filled with smoke, air gray and overcast, the land barren as a moonscape. What happened to all the piñons and junipers? Where was the endless blue sky? The trees must have been clearcut to fuel the coking ovens, I think at first, and the larger ponderosas felled for timbers to shore up mine tunnels. This is probably true. But as indiscriminate as the loss of vegetation is, toxic smoke from the ovens must have played a major role in killing plants and leaving a moonscape. In the photos, smoke hangs over everything, a grim pall. If it killed plants, it certainly didn't spare people working and living around it. Smoke from coke ovens was called "foul gas" for a reason, containing a roll call of poisonous and cancer-causing substances including benzene, toluene, xylene, creosote, and on and on.

I can't imagine working those coking ovens ten hours a day, six days a week, breathing in caustic fumes from the baking coal. When the coke was done, the cokers opened the ovens and sprayed the red-hot coke with water. Imagine all the nasty stuff that spewed out in the steam, straight into their faces, then rose with the breeze to drift off into the environment. After venting the ovens, the cokers raked the smoking, steamy mass of black char into railroad cars, and off it went to make steel. And in their lungs? Residues of chemicals that may have sickened and even killed them.

I don't know the extent of coking's impact on human health, but the photos tell a powerful truth about the impact on nature: the environmental cost of fossil fuels didn't begin with climate change.

There is an even more powerful message: given a chance, the natural world will heal. Those once-bare hills are again blanketed with pines and junipers. The air is so clear and fresh it fills your lungs with joy. And the sky, my wondrous southwestern sky, is blue and infinite as the face of God.

The aqueduct of coking ovens and the related slag heap are dramatic testament to the history of Las Animas County and its century-long love affair with fossil fuel, a love affair shared by pretty much the entire globe. All the mines are closed now and the evidence of coal mining has mostly vanished. But extraction of fossil fuels—coal and now natural gas—is a multilayered story that is still being told.

"We're in the middle of a giant natural gas field." Our neighbor Robin indicates the land around us with a sweep of her arm. We're on the cabin deck sipping margaritas. It's a beautiful summer evening, the soft air buzzing with hummingbirds and smelling of pine. But the conversation is serious. Development of gas wells is gearing up in our neighborhood and surrounding hills as far as we can see, and it's the main topic among the neighbors.

Robin knows gas fields. She and her husband are independent contractors who used to be in the oil and gas business, hiring out their drill rig to do seismic borehole drilling for oil and gas exploration.

"What?" I'm mystified, so she patiently explains boreholes are drilled 60 to 120 feet deep, then a dynamite charge is dropped in and the resulting sound waves recorded to get an audio picture of the underground strata, to see whether they are likely sources for fossil fuels. Now I get it, kinda like bats echolocating! Umm, yeah.

After buying their piece of the wild near us in nearby Madrid Canyon, Robin and her husband made a 180-degree switch in the kind of drilling they did. Now they drill geothermal wells—a "green" energy source—for heating and air conditioning schools and hospitals. Rock on, Robin and Kevin!

I know Robin loves this land—she and Kevin make a near-nightly tour of our association roads in their jeep to check on things—and I guess she no longer wanted to be part of the fossil fuel industry, no matter how tiny her role.

The Raton Basin in which our land lies stretches across some four thousand square miles of southeastern Colorado and northeastern New Mexico. Beneath the ground are abundant fossil fuels, the extraction of which has driven the economy in this area for over one hundred years. And it's still going strong.

Geologically, the Basin is a big underground bowl. The Sangre de Cristo Mountains form its western side, the Wet Mountains its north. Three geologic formations—the Apishapa, Las Animas, and Sierra Grande arches—frame it on the northeast, east, and southeast. The "cereal" within this bowl is a vast accumulation of carbon materials, nutritious food for the voracious appetite of our energy-hungry world. That carbon cereal—coal and natural gas—is what the author Thom Hartmann brilliantly refers to in the title of his 1998 book as *The Last Hours of Ancient Sunlight.*

Talk about boiling things down to their essence. Coal and natural gas really are the reservoirs of ancient sunlight, the energy emitted by an even ancient-er star we call the sun. Millions of years ago, green plants (and microorganisms) captured that energy through photosynthesis and converted it into, well, more plant stuff. The accumulation of a million years of dead plant matter was gradually buried by soil and water. When plant (and animal) tissue breaks down, what is left is carbon, the key element in organic tissue. Over millions of years, immense pressure and heat compressed the ancients' carbon-based cells down to their organic essence, transforming them into coal, oil, and natural gas. Voila!

All that transformed carbon lay in the ground for millions more years as other rock built up atop it. Dug from the ground and burned for energy to power the nineteenth, twentieth, and twenty-first centuries, this ancient sunlight literally transformed life—for the better for humans, not so great for the rest of animal life on Earth, which has literally been losing ground since.

Though area coal mines were largely played out by the mid-twentieth century, another fossil fuel was hiding underground waiting to be recognized. Like coal's shyer sister, coal-bed methane was hovering in the wings all dressed up in (highly flammable) ball gown and tiara. Coal-bed methane is natural gas that formed in conjunction with coal over millions of years. As flashy big sister coal was developing into a coal bed, shy sister methane was insinuating herself into nooks and crannies of the coal seam and in the surrounding sandstone. Until the end of the twentieth century, coal-bed methane wasn't considered a valuable resource at all, but a terrible danger. Methane seeped into mine caverns and shafts, silent and unseen. One spark from an iron pick on a rock could cause a massive, deadly explosion. Odorless and invisible, its very presence could mean death. Breathing methane leads to carbon monoxide poisoning. Utility companies that provide natural gas to homes mix a rotten egg scent into methane to warn homeowners of a gas leak before it kills everyone. In coal mines, the early warning system was a cage of canaries. If the little birds dropped lifeless to the cage floor, everybody knew the mine air was filling with deadly methane. Hopefully the miners could get out of the mine before they, too, ended up dead on the floor or buried in an explosion.

For decades mines vented dangerous methane to the outside, where it dissipated in the atmosphere. In 1984 two million cubic feet of methane per day was being ventilated from just three mines in the western part of the Raton Basin. But in the last twenty-five years or so, coal-bed methane has been recognized as a valuable fuel in its own right, and what was once a massive coal field all around this area is now an active natural gas field.

But you can't dig out gas and haul it from the ground in coal carts. Methane is held in its nooks and crannies by the pressure of ground water. It is, however, lighter than air, so pump out the water that's pinning it in the ground and it will rise on its own to the surface. To get at the coal-bed methane, extraction companies use the controversial practice of hydraulic fracturing, "fracking," to crack the rock of the nooks and crannies. Then they pump out the water, capture the rising methane, and transport it away in miles-long pipelines.

Fracking is its own complex issue, so I'll just say that in an arid land, the practice of pumping fracking chemicals into precious groundwater and then pumping out that chemical-tainted water and dumping it at the surface, where it can infiltrate other ground and surface water, invites a lot of conflict. Sometimes that scarce groundwater is then injected deep underground, a colossal waste in this arid region.

Natural gas (methane) is considered a better, "cleaner," option to burning coal or oil. It's still a fossil fuel, but modern power plants that use natural gas release into the atmosphere half the carbon released by coal-fired power plants. From a climate change mitigation standpoint, natural gas is a transition fuel, like methadone is a transition drug for people getting off heroin. Though it's definitely intruded on our piece of the wild, I am OK with natural gas as long as the world truly transitions to renewables like solar and wind and breaks our addiction to fossil fuels.

One morning a massive drill rig rumbles down our road. The reality of living in a fossil fuel field has arrived on our doorstep.

When we bought our land, the salesman advised us that the mineral rights, including oil and gas, are not part of the purchase. They are, in the parlance of Colorado minerals law, "severed" from ownership of the surface land and had been sold off separately. Then there's the tiny little fact that under state law, mineral rights owners have nearly unfettered ability to remove the minerals, even over the objection of surface owners. Prioritizing mineral rights over surface dates from the state's early gold and silver mining days and the legislature's eagerness to encourage development of natural resources.

"It won't be a problem," the salesman tells us, and everyone else looking to buy land. His tone is light, almost dismissive. "If they do develop the minerals, which they might never"—he emphasizes this—"it will just mean something the size of a Tuff Shed on your land." No big deal. We won't even notice.

Right. We quickly learn that the energy company holding the lease is already at work. They hold a meeting with our landowner association at the local Holiday Inn, which we naively think might give us a say in how wells are developed on our land. But we soon see that's not gonna happen. They are just checking a "met with surface owners" box. At least they offer us refreshments. Over plastic cups of Diet Coke, the company manager and I have a brief chat. "Under Colorado law," he says to me, "we can do whatever we want." The obvious next phrase hangs unsaid but very clear, "and you can't do anything about it." Which, we also learn, is pretty close to the truth, though not entirely. His comment seemed arrogant to me, an attitude that would lead to trouble. And it did.

In the next months, work crews arrive to begin cutting roads and drilling gas wells. One morning, the son of a neighbor in nearby Widow Woman Canyon angrily confronts a drilling crew when they drive onto his land. There is much yelling and fussing, shaking of fists in faces, shouted threats. No one backs down, and the gas company people call the sheriff. The neighbor's son gets arrested. It's probably just luck that nobody ended up shot.

The gas company won that battle, but it was not a good look. Landowners grow more agitated, have more meetings, confront more drill crews, hire lawyers. All the action is taking place in nearby canyons, but we fear it will arrive in ours soon.

But then there is suddenly a new gas development company on the scene. They've taken over the lease from the one with the arrogant manager. Was the fuss and resistance just too much trouble for the first company? I wonder whether they had only dealt with surface owners who were also mineral rights holders. If you earn royalties on gas and oil extracted on your land, you're a lot more willing to tolerate the wells and destruction.

Whatever the backstory, the new company has a better game. To avoid more conflicts with landowners, they purchase all the unsold lots in the canyons. It is on these they will drill future wells, including "slant wells," which somehow go sideways to get at gas on adjacent land (verra sneaky!). One of those lots is next to us, and eventually we get The Letter. A gas well is to be drilled on the land to our north, on the flank of Rattlesnake Ridge.

First a crew arrives with a giant bulldozer and spends days cutting a road up one side of the Ridge to what will soon be the drill site. They clear a huge area, cutting trees and bulldozing the slope into a flat, open dirt patch. Then the massive drill rig rumbles down our road, like an enormous, noisy brachiosaurus. For a couple of days we hear *ka-whump, ka-whump, ka-whump* as it drills down through rock. I can see the brachiosaurus's long neck poking above the line of trees, and at night, the glow of arc lights floods the night sky and washes out the stars.

Then they are done. The brachiosaurus and its attendants trundle away. We walk over to inspect and find a wide, barren dirt patch of well pad with a collection of pipes coming out of the ground and connecting to more pipes, with all manner of gauges, as if they've come out of Captain Nemo's *Nautilus*. And indeed, there is a tiny hut about the size of a Tuff Shed.

We can't see the well from the cabin, but it's not that far away and we can certainly hear it, a constant low-level noise like a Greyhound bus idling. It ingeniously uses the gas it's collecting to power the well, but it's like having a gas generator running 24-7.

I can usually tune out the constant low drone, unless the wind is blowing our direction, but Rick hates it. Plus, an employee in a pickup roars up the gravel access road to the well once a day to record readings from the gauges. We know the truck is here before we hear anything, because the dogs stare suddenly toward the road and erupt in barking.

Can we live with this? Should we sell? Can we part with this cabin we built with our own hands?

We start coming down less often. But after a few years, there is a new development. The gas company electrifies the wells! The Grey-hound bus is gone, replaced with . . . quiet. A few more years and they install satellite-relay technology for remote data collection. Bye-bye, daily gauge-reading guy!

We've lived next to a gas well now for more than twenty years, and we've learned that we *can* live with it. But this change to our piece of the wild is just one pixel in the big picture of what's happening all around.

———————

Labor Day weekend, September 1995. We are brand-new landowners! We set up our backpacking tent in the shade of ponderosas just uphill from the little pond on our new thirty-seven-acre piece of the wild. Bird nerd that I am, I have brought a hummingbird feeder and a jar of sugar water. I fill the feeder and hang it from the lowest branch of the closest ponderosa, within easy view of my lawn chair. Within an hour, a male broad-tail buzzes in to feed. I am happy.

At night the air is soft, fragrant with pine and night-blooming flow-ers. A coyote family sings from down where the arroyo deepens into Toro Canyon. Nighthawks hunt above us, high-flying shadows winging to a tango beat—flap, flap, flutter, flutter, flap. All around, the hills are dark except for two lonely lights on the distant slope of Long Canyon, perhaps three miles off. We see no other lights, no houses, no roads. We have our little piece of the wild and it is good.

That was then.

In the twenty-five years since, things have changed. That we are part of that change I can't argue, but our footprint—cabin, driveway, Tuff Shed—is small compared to what the gas companies have wrought. With all its environmental downside, at least much of the despoiling of the land during coal mining took place beneath the ground. Not so with natural gas extraction. As development accelerates, gas company roads begin to stripe the slopes around us like lashes from a whip.

Driving local roads these days, we pass many side roads leading to gas wells. Sometimes we hike past well pads, but the view from the ground tells only a tiny part of the story. On Google Earth we see the true extent of gas development around us. From fifteen miles above Earth, through the satellite's cold eye, a pale network of wells and their access roads spreads across the hills and over the ridges for miles around the cabin. From space, the well sites look like commas, each wide well pad trailing a curving tail of road. I try to count the number of commas within a five-mile radius of the cabin—at least 264 before I stop, defeated. Hundreds of commas, all connected.

The well pads are each about one thousand square feet, a patch of barren, flat dirt from which the trees, plants, and every natural thing in the way has been bulldozed into a tortured heap at the edge. A small shed and some Captain Nemo's *Nautilus* piping are left in their place. The access roads vary, but I judge the one closest to the cabin to be about six hundred feet long and fifteen feet wide. The math isn't hard. Well pad plus access road account for nineteen thousand square feet of cleared land per well. Times 264 wells (at least). Converted to acres. Hit equal and . . . within this five-mile radius, 115 acres of forest, of wildlife habitat, of carbon-uptaking, oxygen-emitting green vegetation, have been lost, cleared to brown dirt. But not in one piece. Instead, they are spread across the landscape like a web, fragmenting habitat and creating barriers to wildlife movement.

I once started herb seeds under a bell-shaped glass terrarium. It was warm and moist within the bell jar, ideal for plant growth. After a week I noticed ghostly filaments growing between the plant cups, scribing a network of translucent threads. They were the hyphae of a fungus, from the Greek word for web, *hypha*. Long, thread-like cells, connecting end to end, ever branching, spreading across the soil of the plant cups, draping from emerging herb shoot to emerging herb shoot, eventually completely cloaking them in a fungal embrace.

Likewise, the web of gas wells and roads spreads across the landscape like hyphae. Along ridges, up hills, down arroyos. And like fungi, the wells feed on the carbon nutrients of the substrate beneath them. And like fungi, their hyphae will continue to spread across the landscape, drawing the carbon from it until it is all gone.

I hike 150 feet up Rattlesnake Ridge to the rocky outcrop that offers a view across our land to the hills beyond toward the east and south. It's late May and the claret cup cacti that poke from the base of the rocks are in spectacular bloom, offering their waxy, magenta flowers as a gift just for me.

A rock ledge at the top beckons, and I settle myself on a sandstone seat that, like the cactus flowers, is also just for me. It's one of those days I never grow tired of—endless sky I can almost stroke with my hand, pony herds of pine and juniper galloping up one rise and down another, greening plants stepping out in spring finery to stand upright in the warm air after a long winter, opening their flowers like umbrellas to whoever is looking, be it butterfly, hummingbird, or me.

But there is much within this view that isn't lovely. From here I see the many lash marks of roads cut through the forest to gas wells. And on the flank of Rattlesnake Ridge, the dirt pad of the well below me, also cleared in the forest like an open wound.

There is an irony involved here that I think of often. I and my family and my entire community and society have benefitted enormously from a fossil-fuel-based world. We are appalled at the environmental costs—past, present, and future—but we have been the beneficiaries. Like most Americans, I live a life of physical comfort greatly enhanced by fossil fuels. I am warm in my cabin, and my house, on ten-below-zero winter nights because of natural gas. I purchase an infinite variety of goods, many made from petroleum products, and have them shipped from around the world using transportation powered by fossil fuels. I have freedom and mobility because of my gas-powered car. The very fact I can enjoy a weekend cabin a three-hour drive from our primary home is because of gasoline.

We're in a bubble of human history, when great numbers of people can live lives of material comfort, even excess, in a way previously granted only to society's privileged—royalty and aristocrats. Fossil fuels have allowed millions of us to become "the privileged." But these fossil-fuel-driven glory years have come with a heavy price.

It's been a great ride. But though we have benefitted from fossil fuels, it doesn't mean we can't now support moving away from them. We know there are serious, long-term costs and consequences to continuing our rapacious consumption of carbon. Replacing fossil fuels with clean energy sources doesn't mean the comforts of modern life need to disappear.

Coal fueled the Industrial Revolution, allowed the vast development of our planet, the explosion of material abundance, and the technological advances of "modern" life. Natural gas is carrying on the tradition. It took less than two centuries for much of that ancient sunlight to be pulled from the ground and burned to release its energy. Obviously a fuel that takes millions of years to form but just a dozen decades to deplete is not a renewable resource. The harvesting of coal and its sibling fuels—oil and natural gas—led to a massive footprint left on the Earth by just one species, *Homo sapiens*. The party was great while it lasted, but now a party crasher known as climate change is here, presenting the world a bill for the celebration. And our land is already paying the price.

13

SNOWBOUND

Blizzards are predicted to become more intense in the face
of climate change, despite shorter winters and rising global
temperatures.
—"Maybe It's Cold Outside,"
  NationalGeographic.org, June 18, 2020

CABIN JOURNAL—DECEMBER 26, 2006, TO JANUARY 2, 2007: *We
arrive the day after Christmas to find perhaps six inches of residual
snow in the meadows from the big Christmas storm, much deeper in the trees
and arroyos. Trinidad had twenty-four inches, but it has mostly melted and
packed down in the meadows and on the roads. But Thursday morning the
28th, we awaken to another heavy snowstorm. It snows all day and night,
continues all day and night Friday and all day till sunset Saturday. We
watch the meadows fill with snow, until it completely covers the stone firepit
and log seats, the smaller trees, and is up to the window sills of the cabin.*

I have been through many, many Colorado snowfalls, some of them
"just a skiff," others measured in feet. But the after-Christmas snow of
2006 was the biggest I have ever seen all in one dump. By the time it
ended, nearly four feet of snow lay over the cabin and everything around it.

Snow speaks to something deep within us. Falling silently, its motion
is somehow still. Snow is calming, mesmerizing. It falls gently yet builds
up mightily. It erases hard edges, layers the known with mystery. It puri-
fies the harsh, transforms the ugly. It fills the barren. Snow is powerful

and unrelenting, but its power is without malice. Snow has no ego or intention. It just is.

Like Robert Frost pausing on a snowy evening to watch the woods fill up with snow, I gaze from the cabin across our meadows and forests as the snow falls, and falls, and falls. Its draw is nearly irresistible. I want to go out in it, lay first tracks by boot or ski or snowshoe, embrace its cold softness. But you can't embrace snow and also preserve its stoic beauty. It changes at your touch, transforming to water, melting away. I don't want to change this snow. It is perfect and unmarked. Newborn.

We stay indoors at first, tucked snugly in our warm, glowing cabin. We don't have Wi-Fi or television. We sit cross-legged on the rug in front of our cast-iron heating stove, shifting occasionally like toasting marshmallows so we're evenly browned on all sides. As the silent snow builds up outside, burying us, it feels as if we are the only people in the world.

"Let's play a game," Olivia says, pulling Dogopoly from the game cabinet (Monopoly, but with Chihuahua and Great Dane instead of Mediterranean Avenue and Boardwalk). She plays pretty expertly for an eight-year-old, easily beating me and just edging our Rick.

We enjoy the game, but the snow is beckoning and we're all eager to get out in it. We put away Dogopoly, strap on our snowshoes, and head across the meadow toward Toro Canyon. Our tracks stretch behind us to the cabin door, an ephemeral tether slowly erased by the falling snow.

The woods are hushed, all sound softened. Snow-flocked pines rise up the mountain slopes all around us like choirs of white-robed angels. Bare rocks poke from vertical cliffs too steep to hold snow. We see almost no animal tracks, they have been filled in by snow, but a few birds move among the tree branches—chickadees, scrub jays, flickers, ravens. For wildlife a winter snow is just another day at the office.

In the secret meadow, we build a snowgirl. She has juniper-frond hair and pine-cone eyes. We add sticks for arms—one bends up at the wrist, the other curls down to her hip. A snowgirl with attitude, like Olivia!

The snow keeps falling. We pass the days reading the latest Harry Potter book out loud, doing a jigsaw puzzle, playing with Olivia's American Girl dolls—Samantha and Marisol. We bring out paints and paper, scissors, tape, glue, and decorate the cabin with our creations. Snow days are good arts-and-crafts days. And good days to bake. We up our daily calories with brownies and gingerbread.

But we can't bear to stay indoors for long. There have been a lot of snowstorms at the cabin, sometimes leaving ten inches of snow, but never anything this big. We bundle up and wade out into the meadow through snow that is now waist-deep. We swan dive into the soft powder, landing with a gentle *whump!* I grab up handfuls and throw snowballs at Olivia and Rick. They retaliate, but the hard, fresh powder is too dry to hold together and disintegrates into sprays of snow. We chase each other, moving in slo-mo through the deep white stuff as we dump snow on each other's heads, laughing and panting foggy puffs with each breath. Rick and I take Olivia by her arms and legs, swing her back and forth—One! Two! Threeee!—and pitch her squealing into the cold, white pillow of snow. The dogs surge through the snow after us like icebreaking Arctic ships—Jasper is big enough to bull his way through, but Rosie leaps and bounds like a winter ermine.

There are many adventures in this winter landscape. Rick and Olivia dig a snow cave just big enough to hold both of them. They crawl in and sit cross-legged like little snow bears, playing cards and drinking cocoa by the light of a votive candle, which gives just enough heat to warm the small space without melting it.

Rick snowshoes up to the neighbor's house atop Rattlesnake Ridge and knocks on the door in a blizzard version of "Can I borrow a cup of sugar?" though Rick is seeking news not baking supplies. The neighbor is stunned that out of the blizzard, in the middle of nowhere, someone is knocking on his door. Unlike us, he has satellite TV and updates Rick on the storm. It's a massive snow dump all along the Front Range but especially fierce in southeastern Colorado, following on a two-foot dump a week earlier.

"Daddy, you were like Mr. Edwards in *Little House on the Prairie*," Olivia says, reminding us of the scene in the book when the distant neighbor appears out of a snowstorm to share Christmas with the Ingalls family. And we do feel a lot like Laura Ingalls Wilder and her family, alone in the wilderness, cut off from the world but all together safe and warm in our cabin, having so much fun with simple things. Though after three days of nonstop snow, the book in the Little House series it makes me think of most is *The Long Winter*.

The snow stops Saturday evening. The sky clears to a spectacle of stars, sequins pinned against a sky of blackest black. Orion stands in the southern sky, as if he had climbed up Raton Mesa and into the heavens to reassure us that all is well. A bright moon rises, lighting the sea of snow until the world is a cold, glowing blue, glittering with diamonds.

My thoughts turn now to getting home. Tomorrow is New Year's Eve. We had planned to drive back to Castle Rock the day after, on New Year's Day. Olivia is to go back to school and Rick back to work. But we are surrounded by four feet of snow. How long might we really be here? Will the power go out? Might we run out of water? Could we run out of food?

I joke that the dogs are looking worried, in case they end up like the sled dogs in every failed Arctic exploration—in the stewpot. Not to worry, puppies! We have bags of rice and cans of beans and tuna in the food cupboard. Dinner might be dull, but we won't starve (or have to barbecue the dogs). I'm most worried about running out of coffee. Olivia frets as the supply of chocolate gets low.

Rick goes out to start the Tahoe, which we've done regularly throughout the storm to keep the battery charged. The snow is up to the windows of the car. I watch from the cabin as he brushes off a mound of snow, making sure the tailpipe is clear, and gets in. The engine flares to life and after he's let it warm up, he puts it in gear and the Tahoe inches forward. But only inches. I see the car jerk slightly as he puts it into reverse. It creeps backward just a few inches. Forward again, reverse again.

In a few minutes, Rick comes back inside, stamping snow off his boots and brushing it off his snow pants. His expression is grim. "There's no way we can even get to the gate, even in four-wheel drive. Just the little bit I moved it, the snow was packing up underneath and the Tahoe was sinking into the snow. We'd be stuck before we got ten feet." We look at each other as reality sets in.

We are snowbound.

"If you don't like the weather, wait five minutes and it'll change." It's a common reaction to Colorado weather, and it's pretty accurate. A sixty-degree day can plummet forty degrees and end with ten inches of snow. A sunny afternoon can turn suddenly into a frigid hailstorm, then in ten minutes be bright and sunny again. And an average winter day can transform into a blizzard that brings one, two, even three feet of snow.

Giant snowstorms and swings in the weather have been a part of life in Colorado since long before greenhouse-gas buildup in the atmosphere began changing the climate. The biggest Denver snowstorm on record was back in 1913, when 45.7 inches of snow fell between December 1 and 5. In fact, eighteen of the biggest snowstorms in Denver history occurred before the twenty-first century, including three in the nineteenth century. The mountains, of course, regularly get big dumps of snow. It's news when they don't.

So the Big Snow that left us snowbound in 2006 was not surprising, though its forty-plus inches in three days came close to Trinidad's entire average annual snowfall of 63.7 inches. Was the Big Snow a product of climate change? That's a great big maybe. It's difficult to tie specific storm events to a long-term cause like climate change, though scientists found that abnormally warm surface water in the Atlantic, due at least in part to climate change, contributed to major blizzards on the East Coast in 2010.

What our Big Snow *was* is a glimpse of the extreme weather—blizzards, tornadoes, flooding, severe heat, and more—the entire world is going to face as climate change gets worse. And an example of why the term *global warming* is misleading and has generally been replaced by *climate change* to describe what warming of the globe leads to. The globe is indeed warming, but that contributes to extreme weather events that are sometimes cold.

Extreme Weather. It sounds like a video game for weather geeks. All fun and safe sitting at a game console, not so fun if you're going through it. But the world will have to get used to, and hopefully be more prepared for, big weather events. Numerous extreme blizzards have hit all over the United States since our Big Snow in 2006. Some were given dramatic names. Snowpocalypse hit the Northwest in 2008. A second Snowpocalypse, rated a Category 4 "crippling" blizzard, nailed the Northeast in 2009. Snowmageddon clobbered the Northeast again the next year, 2010, with thirty-eight inches of snow and at least forty-one deaths. Snowzilla, a Category 5 "extreme" blizzard, moved from the mid-Atlantic states through the Northeast in 2016, killing fifty-five people and causing up to $3 billion in damage and losses.

Even official meteorology terms can't stay dull with all the crazy weather events. A major blizzard hit south of Denver in March 2019. Meteorologists described it as a *bomb cyclone,* a term I'd never heard before, which involves high winds and a cold air mass colliding with a warm air mass. The barometric pressure dropped twenty-three millibars the day it hit, which is apparently a stunning event. It was the lowest pressure ever recorded in Colorado. With big snow, big cold, and big winds, the bomb cyclone was described as the most intense storm along the Front Range since records began. I don't know whether it was linked to climate change, but it definitely ranks as never-before-seen Extreme Weather.

The obvious question is: Shouldn't a warmer, drier climate mean less snow and smaller snowstorms? Climatologists say no. Though winters are getting shorter and global temperatures have increased (1.76 degrees Fahrenheit compared to the twentieth-century average, according to NOAA), climatologists predict extreme winter weather events are going to get only *more* extreme.

The simple explanation lies with the colorful graphic we all studied in elementary school: the Water Cycle, with arrows going from the ground and ocean up into the sky (evaporation) to form clouds (condensation), and from clouds down to the ground (precipitation). Warmer air temperatures and warmer oceans mean greater evaporation of moisture into the atmosphere. That moisture forms fatter, heavier clouds. When the clouds can't hold all that moisture any more, they let it fly, as either rain or snow. Bigger dumps of moisture mean bigger rainstorms and snowfalls. What might have been an average snowstorm in earlier times becomes Snowmageddon.

Of course, climate and weather, and all Earth's processes that create them, are far from simple. For various reasons, the Arctic is warming faster than other parts of the globe, which pushes the jet stream farther south, taking cold Arctic air with it (the Arctic may be warming, but it's still really cold). Where that cold air goes, extreme snow events can result.

That doesn't mean we'll end up with more total precipitation in a year. Instead of a lot of small storms in a year, there will be fewer storms but they'll be much more intense. Result? Extreme Weather.

Over the holiday season of 2006, the Colorado Front Range was the lucky winner of the snow lottery jackpot. Think PowerSnowball or Mega Millions-of-Snowflakes. First came the pre-Christmas blizzard of December 20 to 21, which dumped as much as three feet along the

Front Range and eastern Colorado. Wind piled drifts up to five feet deep (snowstorm + wind = blizzard). The entire eastern part of the state was closed down in terrible blizzard conditions.

We dug out from that snowstorm enough to get down to the cabin the day after Christmas. Then on December 28 came Holiday Blizzard Part Deux. The Big Snow that left us snowbound hit farther south than the pre-Christmas storm and was even bigger, dumping three to four feet of snow across the southern Front Range and the southeastern plains. Thousands of people were stranded, and an estimated ten to fifteen thousand cattle died of cold, exposure, and starvation.

And we were snowed in at the cabin.

Our driveway is 0.28 miles long from our gate to the cabin. A Chevy Tahoe is seventy-nine inches wide. The snow has accumulated about forty inches deep on the driveway. That means there are 32,443 cubic feet of snow that need to be shoveled, plowed, or compacted to make a track to get us out of here. Or, doing more math, 1,201 cubic yards of snow to move. OK, that sounds better.

We are snowed in, with no snow-cavalry on the way to save us, which means we need to start saving ourselves. We have three people to shovel snow, but one of them is only eight, so she is pretty much done after three or four shovel loads. So Rick and I start shoveling. Full disclosure: Rick did most of it, spending days digging away at the driveway until it is a sunken trough, still knee deep in snow, all the way to the gate. But it's still too deep and soft to drive on. Shoveling ourselves out was a nice idea, for lack of any others, but, of course, it was impossible to shovel that much snow—1,201 cubic yards!—so we turn to Plan B. We will snowshoe up and down the driveway over and over, hopefully packing the snow enough to give the Tahoe a base to drive on with the help of four-wheel drive.

Even if the car makes it to the gate, however, we can't go any-where if the roads aren't plowed. Our community's agreement with the natural gas company lets them use community roads, but they are responsible for maintaining them. But what if they make other roads a priority and leave our little stretch of dead-end road until last? It could be days before they clear our access road. So Rick makes a sign on a piece of heavy cardboard—TWO FAMILIES SNOWED IN—with a big arrow pointing down our road (the second family is the neighbor up on Rattlesnake Ridge). Off he snowshoes across the meadow, down and up Toro Canyon, to the junction where our road forks off the primary road. He tacks the sign to a two-by-four post and sticks it in the snow as prominently as he can. Then it's back to the cabin, to keep tromping the driveway and hoping.

The next morning, we hear snowplows in the distance, but they aren't coming closer. By afternoon, with our ears tuned for large vehicles coming our way, we go for a snowshoe adventure in the forest. Snow lies deep amid the trees, layered on the branches like white frosting. The tracks of animals now tell stories of life emerging from shelter to go about business—messy troughs left by a bounding mule deer, the twin zipper tracks of a pinyon mouse that is light enough to skitter across the snow without sinking.

We check on our snowgirl. She still stands with hand-on-hip attitude, inches of new snow built up on her juniper-frond hair. It's about 3:00 PM, but the winter sun lies low above Elk Ridge, casting long, lean shadows across the secret meadow. Evening comes early here in midwinter, bounded as we are by ridges on three sides. Suddenly my ears perk up like Jasper hearing a coyote. The grind of a large vehicle. It's hard to judge the distance, but it's close. Please, oh please, let it be a snowplow coming our way.

We race back to the cabin as fast as we can on snowshoes through deep snow. Rick heads for our gate to flag the plow driver. Olivia and I rush into the cabin and start throwing things into duffle bags. The sun is setting, and we need to be ready to jump in the Tahoe and head home if we get plowed out.

Rick's sign worked! The snow-cavalry is coming to rescue us, trundling slowly along the road. It's actually a road grader, not just a snowplow, with giant tires that aren't daunted by Extreme Weather. The driver's job is to plow the roads (which are all just gravel), but bless his kind soul, he turns in through our gate and plows all the way up to where the Tahoe sits in the snow like a marooned bull bison. Take that, 1,201 cubic yards of snow!

The driver leans out his cab window and grins. "Saw your sign. Guess you folks could use some help." We chat a few minutes, and he is very friendly, updating us on the state of the roads all around the area and up Interstate 25 north to Denver. Not great, he says, but the highways are clear and driving should be fine once we're off the unpaved roads. We offer him twenty dollars in thanks, but he shakes his head. "Glad to help you." His kindness touches my heart, reminding me that most people are good—at least that's what I like to think.

"Do you want something to eat or drink? We have brownies," I ask. He endears himself to me even more by accepting both a brownie and a Diet Pepsi. My kind of calorie watcher. We watch as he edges into the meadow to turn around, maneuvering the grader forward and back, forward and back until its nose is pointed back downhill toward the gate. Then with a wave of his hand he is off, trundling out to plow more roads. I wonder whether there are more stranded neighbors he will save. Very likely.

Rick starts digging out the Tahoe, which now has a ridge of plowed snow in front of it, and Olivia and I go back inside to finish packing. In ten minutes we are in the Tahoe ready to roll.

The evening hovers in a blue twilight as we start toward the gate. Driving these roads in the dark and after a dump of snow is far from safe, and if we get stuck, it could be an ordeal to get out, not to mention dangerous if we are stuck in the car all night in below-freezing temperatures. And we do immediately get stuck, twice, just on the driveway, but each time, Rick wedges a mat under the tires for traction and we keep going. The gate is another challenge. It is a metal ranch gate, with two halves that swing together to meet in the middle. Both sides are propped open, but they're so buried in snow we have to dig them free before we can swing them closed and chain and lock the gate.

Finally, we are on the road home, making the drive north on Interstate 25 in the dark. In my mind I thank the plow driver over and over. There is no way we could have made it out on our own. He is paid by the gas company, and I laugh to realize it is just one more way we benefit from fossil fuels.

The moon illuminates an endless sea of white, lapping all around to far-off hills. We drive in silence, Olivia asleep with the dogs cuddled next to her, the tires on the snowy road humming a wordless song that carries us home through a harsh but beautiful land.

We arrived back home in Castle Rock on January 2 by 9:00 PM to find only about eight inches of snow. Castle Rock had dodged the snow bombshell that hit the cabin full force.

But the Holiday Blizzards weren't done delivering gifts. Olivia and Rick went back to school and work on January 3, but two days later a third blizzard hit. It delivered another foot of snow.

Between blizzards, tornados, floods, and other events, Extreme Weather in the last fifteen to twenty years has had an enormous economic cost. The triple-whammy blizzards of December 2006 to January 2007 likely cost $100 million or more in losses and damages, not to mention thousands of dead cattle and several human deaths.

The Big Snow that left us snowbound at the cabin may or may not have been caused, or made worse, by climate change. But it greatly affected people, wildlife, livestock, and livelihoods. As climate change leads to more extreme weather events like the Big Snow, the costs—financial, emotional, and mental—are going to be huge.

We don't return to the cabin until Easter weekend in early April. When we left, the meadows lay sleeping beneath deep snow. Now bluebirds busily investigate the nest boxes, the grass is greening, and early wildflowers like sand lilies and yellow violets sprinkle color across the meadow. The land is reaping dividends from the Big Snow. In the way of the natural world, winter is forgotten and life moves forward, always forward, into the next season.

And for us, more surprises are waiting.

14

A MOUSE IN THE HOUSE

Climate change has adversely affected physical health of people globally. . . . The incidence of vector-borne diseases has increased from range expansion and/or increased reproduction of disease vectors.
—*Climate Change 2022: Impacts, Adaptation and Vulnerability, Summary for Policymakers*, Intergovernmental Panel on Climate Change, February 28, 2022

CABIN JOURNAL—APRIL 2007: *We arrive at 7:30 PM Saturday evening, the night before Easter. We haven't been at the cabin since being snowbound after Christmas and barely making it out. Now, the land will be greening, and the bluebirds on the hunt for nest sites. We open the cabin door and turn on the lights, ready to relax. But something isn't right. The cabin is overrun with intruders.*

We don't immediately realize we have home invaders. No windows are jimmied, all the doors are secure. The intruders are much too sneaky, and small, to use such obvious entries.

It's well past sunset when we arrive at the cabin, and we hustle to carry our duffels and bags of food from the car by the light of the moon and the weak glow cast by the outside light above the kitchen door. I open the food pantry to put away the dry food and notice pale crumbs on one of the shelves. Then I find grains of rice spilled on the floor. An apprehensive light flickers on in my brain. *Is this what I think it is?!*

Sure enough, an unopened bag of rice has a mouse-sized hole chewed in one end. A box of shredded wheat is likewise despoiled—the source of the pale crumbs. I look closer at all the shelves. Canned food and hard containers are fine, but there is damage to every cardboard box and bag of food that mice teeth can gnaw into. Worst of all, telltale black mouse droppings, the size of grains of rice, are everywhere.

"Rick," I call out, "we have a problem."

My first job out of college was as a restaurant inspector for the Denver health department, a desperate I-need-a-job choice after a last-minute decision not to go to grad school. One thing I learned from three and a half years inspecting restaurants was how to recognize a mouse infestation.

I grab a flashlight (the health inspector's trusty friend), and we begin an inspection of the cabin. There are droppings in the corners and along the walls. That's enough for us to put on the paper filter masks we keep on hand for various jobs. This time they're to protect us from hantavirus. Hantavirus pulmonary syndrome is a respiratory illness transmitted in the urine, droppings, and saliva (left on things they've chewed) of native mice of the genus *Peromyscus*, including our local pinyon mice, *Peromyscus truei*. Peromyscans are the common, native deer mice of North America, not the naked-tailed house mouse, *Mus musculus*, which is an invasive species from Europe.

There are six Peromyscan mice species in Colorado, and the organism that causes hantavirus can potentially be carried by any of them. The name *Peromyscus* derives from Greek and translates to something like "booted little mouse," a reference to their pale feet. As a group they're commonly known as white-footed mice or, more familiar in the West, deer mice. These agile little mice got this nickname because they jump really well (like deer), something I've learned several times when I opened a nest box expecting a bird nest only to find a bright-eyed pinyon mouse staring at me, which in the next second vaulted past my head like Simone Biles, landed on a nearby branch, and scampered off.

Since pinyon mice and other Peromyscan species are everywhere in the Southwest, hantavirus is an ever-present concern. It's not common, but it is no trivial disease for those who contract it. Nearly 40 percent of cases end in death. Pinyon mice may be cute and doe eyed, with furry tails and Dumbo the Elephant ears, but they're definite hantavirus carriers. Inhaling household dust that holds the dried residue of their waste carries the risk of a potentially fatal respiratory infection.

The cabin's mouse infestation is a bad one. We find evidence of mice pretty much everywhere—in all the closets, behind the refrigerator, among the pots and pans, in the cabinet under the kitchen sink, in the cupboard where we keep the games. They've chewed the tassels off our woven placemats and begun nesting in the drawer with the kitchen towels, chewing the fabric to make nests. In the cabinet and drawers under the bathroom sink, they've pulled out cotton balls for nest material, pranced over brushes and combs, chewed on the cotton ends of Q-tips, nibbled into paper-wrapped bars of castile soap. They've even run across the beds! I imagine hordes of mice dancing in a conga line through the cabin like college kids on spring break. *Conga! Conga!*

Olivia, just turned nine, is horrified at the thought of mice everywhere, and I can barely persuade her to come inside the cabin. I don't want her exposed to the hantavirus risk, so I carefully fold up and bag the bedspreads in her room, clean up any droppings with spray bleach, and install her and the dogs safely in her room with the door closed and a movie on the DVD player.

It's 8:00 PM, and Rick and I are tired after a long workday and a three-hour drive, but we start in cleaning. We throw away everything in the pantry that isn't in a hard, sealed container, even if it doesn't appear to be chewed on. We clean all the shelves with spray bleach, wipe down the cans and containers. Then we proceed through every cabinet, drawer, closet, and surface in the cabin, spraying with bleach to hold down dust and to disinfect, wiping, throwing away anything chewed on or fouled by mice. Into bag after green Hefty bag go food, chewed rolls of paper towels and toilet paper, linens, towels, leather work gloves. I find the twin grooves left by mouse incisors in weird stuff like candles, sponges, and bars of soap. Pinyon mice definitely aren't finicky in their food choices.

Finally at 1:30 AM, we call it done. We've been cleaning and demousing the cabin for five and a half hours. Olivia long ago fell asleep with the dogs around her. Rick and I are way past exhausted, moving like zombie-droids with wipe cloths. But the cabin is cleaned up, wiped down, and effectively disinfected. Tomorrow we will do our best to figure out how the mice are getting in and plug up those entry points. But for now, in the wee hours, we fall into bed for some sleep, visions of pinyon mice instead of sugar plums dancing in my head.

Our cabin is a sanctuary, a place of retreat. Escaping the busy craziness of daily life for the calm rhythms of the cabin is a gift. In spite of the challenge of 1,201 cubic yards of snow on the driveway during the Big Snow, being snowed in was a magical time we all still treasure. So on this first visit after being snowbound, we were looking forward to another pleasant getaway, this one with bright sun, blue sky, and bluebirds building nests in the boxes. Instead, we found the post–Big Snow mouse invasion and a huge cleanup job.

But why was there such a huge mouse infestation? We never had mice inside the cabin in the nearly eight years since we built it. They've invaded the nest boxes and made themselves at home in the shed, but the cabin has remained mouse free. So why this time?

It's hard not to link the two events—the Big Snow and the influx of rodents. Following the Big Snow came the Big Melt, leading to the mice being Big Flooded out of their dens in or near the ground. Forty-plus inches of melting snow spreads a lot of water across the landscape, pushing small rodents out of the hidey-holes they occupied all winter. So they went looking for new, dry shelter and found their way into the cabin. Pinyon mice begin breeding in the spring, so the imperative to find snug nest sites for rearing babies was an additional motivator. No wonder they appropriated so much "soft stuff" like kitchen towels and cotton balls. The cabin became a Babies "R" Us for pinyon mice.

The mouse invasion was likely the beginning of a mouse population spike. The deep snow held moisture across the land much longer than usual and, as spring began, provided an abundance of riches for growing plants. Though April was just the beginning of the season, the flush of plant growth began to bring a boon of food for small mammals in buds, flowers, and hatching insects. Soon the bloom would produce a banner crop of seeds for pinyon mice and other rodents to feed on.

All of this is business as usual in the natural world. Nature is a complex, interconnected network of causes and effects. Like an intricate version of the children's game crack the whip, precipitation leads to plant growth, which fosters a boom in small mammals, followed by a boom in the animals that feed on them. The journal for that year seems to bear this out, a wonderful spring bloom and, later, what seems to be a lot of predator sightings in a short time period.

CABIN JOURNAL—MAY 2007: *Meadows incredibly lush, green with grass, covered with wildflowers—Indian paintbrush, fleabane, asters, purple, white, and lavender penstemon, sugar bowl, purple and multi-color loco. Banner year for milky vetch . . . broad-tailed and black-chinned hummers, western bluebirds, violet-green swallows, ash-throated flycatchers, black-headed grosbeaks, a big flock of pinyon jays. All the usual suspects!*

JUNE 2007: *Around 11:00 PM I hear a great horned owl hooting, soon answered by a second, slightly higher in pitch. Then they duet. Coyotes howl most evenings, once right down at the old campsite. Under the edge of the porch, curled up tightly, is a baby rattlesnake. We pick it up with a shovel—it rattles a lot at that—and drop it in a bucket, then release it in a remote drainage where we don't hike. A crowd of scrub jays mobs a raptor with a pale cheek stripe—a prairie falcon?*

But what seemed like nature's business as usual at the time now seems like a foreshadowing. Whether or not our Big Snow was the result of climate change, it demonstrated that extreme weather events and increased precipitation—which the world is already seeing as the climate warms—can lead to a boom in rodents and potential exposure to diseases they carry.

Evidence is increasing that climate change will foster a worldwide spread of infectious diseases that are carried by "vectors"—mice, rats, mosquitoes, ticks, fleas, and other animal hosts. Just as birds move their ranges northward as the climate warms, so can these vectors, including cold-sensitive invertebrates like mosquitoes and ticks. They, along with the diseases they carry, won't be limited by cold weather and freezing temperatures, or they will be active for more months of the year than previously. Shorter winter season equals longer disease season.

Lyme disease is the most common vector-borne disease in the United States. The bacteria that causes it is carried by the black-legged tick, whose range is limited by temperature and precipitation. A bite from an infected tick causes flu-like symptoms and fatigue, which if not treated, can lead to nasty complications like arthritis, heart issues, and neurological problems. Luckily for westerners, Lyme disease occurs mainly in the northeastern and southeastern United States. But it is expected to spread with changes in the climate. A 2020 study from Stanford and University of California, Santa Barbara, researchers found a link between warming annual temperatures and the increase of Lyme disease in the northeastern United States. Earlier spring and later fall frost meant a "lengthened transmission season."

Ticks. I have a hard time appreciating their redeeming qualities. They have their role in the ecosystem and all that, but still. They're bloodsuckers. Eeew. And that means they need a host whose blood they shall suck. Any guess who that host for the black-legged tick might be? Our familiar friends the booted little mice of the genus *Peromyscus*.

Mice are actually just one member of the black-legged tick team. Ticks have complex life histories, making use of different hosts in different stages of growth. Adult black-legged ticks attach themselves to deer (leading to another common name, deer tick), dogs, and on occasion, humans. White-footed mice and birds are major hosts for the immature larvae and nymphs. While mice don't move very far, birds are very mobile and wide ranging, meaning they can potentially carry ticks, and therefore Lyme disease, lots of places. Black-legged ticks do not currently inhabit the interior West, and western ticks don't carry the Lyme disease organism. But biological systems are always evolving, and black-legged ticks could expand west or the organism could make the jump to new tick species. As the climate warms and new areas become favorable, we could see a big spread of Lyme disease. Even without a major move westward, researchers project that habitat for the black-legged tick will expand to include most of the eastern United States by 2080. Lyme disease is already moving northward. Cases quadrupled in Maine in the ten years between 2005 and 2015 and are growing in eastern Canada.

For the Mountain West, Lyme disease is not currently a threat and not likely to become one any time soon, if at all. But our mouse invasion at the cabin raised the specter of a different disease that does present a threat in the Southwest.

———————

The year was 1993. Whispers of a mysterious plague began circulating in the Four Corners region of Colorado, Utah, Arizona, and New Mexico. Otherwise-healthy young people on and near the Navajo Reservation in New Mexico came into health clinics with unexplained muscle aches in their legs, hips, back, and shoulders. They hadn't been injured and couldn't pin the pain on muscle strain. They all had fevers and complained of general fatigue. Some patients also had headaches, dizziness, chills, stomach pain, vomiting, and diarrhea.

A week or more after they first felt ill, the patients began coughing and gasping for breath. One said it felt like a tight band around his chest and a pillow over his face. The patients' lungs were filling with fluid, but the doctors had no idea why.

Over two months, thirteen people died from the enigmatic ailment and many more were sick. This new "plague" puzzled medical experts and was so mysterious it was dubbed Sin Nombre virus, Spanish for "without a name."

Eventually the culprit was identified as a novel hantavirus, a family of viruses found worldwide and spread by rodents. Every year about two hundred thousand cases of various kinds of hantavirus are reported globally, though the illness in the American Southwest, eventually named hantavirus cardiopulmonary syndrome (HCPS), was new and different. It remains a regional disease with relatively few cases, though in 2012, hantavirus broke out among visitors to Yosemite National Park.

Why now? That's a natural question about the disease's emergence. Clues to the answer came from an unexpected source.

Though officials and medical experts were baffled, there were some in a very old community who had heard of such a thing before. When Navajo tribal authorities sought help from tribal elders, they were directed to traditional healers. Navajo oral tradition, it turns out, told of earlier deadly outbreaks in 1918 and 1933 that came on quickly with no warning. Disharmony was the root cause the healers cited, leading to "excess." In both of those years, as in 1993, an excess of rain and snowfall had led to a huge crop of piñon pine nuts, which in turn led to an explosion in the local mouse population. The rodents got into hogans where people lived and places where they worked and stored food. Mice were everywhere, increasing the chances humans would come into contact with them and their waste.

It was a key connection. Now that they knew to investigate Sin Nombre as rodent-borne, epidemiologists put together the symptoms and other details and quickly realized they were dealing with a new kind of hantavirus.

The second key was the "excess" rain and snow, which answered the question "Why now?" It turned out 1992 was an El Niño year—a season of increased precipitation due to an El Niño-Southern Oscillation (ENSO), a recurring warming of the ocean surface across the tropical Pacific that had a big impact on North American weather. Cue the game of ecosystem crack the whip—El Niño, increased precip, abundant plant growth, mouse population boom. Hantavirus outbreak.

How bad was it? Imagine you see a mouse scurry across your kitchen floor. Then imagine you see twenty mice scurry across your kitchen floor. All at the same time. El Niño excess, researchers estimated, led to a twentyfold increase in the rodent population!

El Niño roared back again in the winter of 1997–1998 with one of the strongest ENSO events ever recorded. Though hantavirus had been all over the news for years and the public was much better educated about how to avoid it, there were five times as many cases of hantavirus as there were in 1993. Most of the patients reported they had been exposed to deer mice (those pesky Peromyscans) indoors. Just as happened at the cabin in 2007, extreme weather led to extreme mice, who invaded people's homes. In the extreme.

Luckily, none of us contracted hantavirus from the mouse invasion of the cabin. The only thing we suffered was exhaustion from five hours of nonstop cleaning and major gross-out from everything being so mousey and foul. But it was not going to happen to us again, mister, believe you me!

The next day, before we button up the cabin to go home, we check every wall line, corner, and crevice we can, inside the cabin and down in the crawl space, looking for entry points. Mice are practically shapeshifters; with squishy bodies and mobile shoulders and hips, they can squeeze through impossibly tiny spaces. We spray foam sealant in every possible opening, any joint between logs, any space around pipes where they come up through the floor and cabinets. The foam spreads into gaps and expands, drying into a rigid, water-tight material that mice won't chew. To get rid of any invaders still hiding in the cabin or to dispatch any new arrivals, we place poison bait blocks and snap traps around the walls inside the cabin, more bait blocks in the crawl space.

Finally, still exhausted, we lock the doors and head home by midday Sunday. We'd been at the cabin less than twenty-four hours. It felt like forever.

I return a week later to check on things. Practically holding my breath, I unlock the door, swing it open, flip on the light. Please, oh please, be mouse free!

I make a quick tour of the mouse traps. We've caught three mice. A fourth baited trap is untouched and unsprung. Best of all, there is absolutely no new evidence of mice. I dance around the cabin with joy! *Conga! Conga!*

I'm not in general a fan of lethal pest control, preferring to use live traps if it's feasible. Which it isn't at the cabin, since we wouldn't be there to release any live-trapped animals outside. And I am never happy to use poisons. But the 750 square feet of our cabin, out of thirty-seven acres of land, is where I draw the line. I can't do anything about El Niño or a Big Snow or a Big Melt or a Big Mouse Baby Boom. But I can police the cabin, and within this box of logs, no rodent shall tread.

In recent years countless communities in the United States have been devastated by extreme weather events caused or made worse by climate change: massive blizzards, floods, tornadoes, hurricanes, wildfires. The immediate impacts of these events are obvious, but they also have downstream effects that don't show up for weeks or months, or perhaps years. Disease is one of these.

Infectious diseases will likely increase globally with climate change, experts tell us. Lyme disease and hantavirus are far from the worst possibilities. There may be links between a warming climate and the emergence of novel coronaviruses such as the culprit in the COVID-19 pandemic. What if a warming climate allows the Anopheles mosquito, the vector for malaria, to survive in places that have previously been too cold or dry? Cold weather has protected much of North America from "exotic" diseases. But if that bulwark melts away, we could all become vulnerable to illnesses we never dreamed of.

15

MARK OF AN ANCIENT CATACLYSM

This is the story of one terrible day in the history of the
Earth. . . . Vast numbers of highly successful animal and
plant species suddenly disappeared in a mass extinction,
leaving no descendants.
—*T. rex and the Crater of Doom*, Walter Alvarez, 1997

Climatic change is the connection between the impact and
the extinction: the impact upset normal climate, with long-
term effects that lasted much longer than the immediate and
direct consequences of the impact.
—"The K-T Extinction," from *History of Life*,
   Richard Cowen, 2000

CABIN JOURNAL—OCTOBER 2019: *We stop at the cliff face along the
county road where the rock strata are exposed. Tucked atop the coal
seam, below a deep layer of sandstone, is a narrow white line, a chalky
layer about one inch thick.*

I've driven past this exposed cliff face a thousand times without
noticing the thin white line of rock, but now that I know what to
look for, it jumps out at me. It has an awkward name—the K-Pg
Boundary—but this white line is testament to one of the most significant events in the history of Earth: the fifth extinction.

"It's along here," I say to Rick after we cross the Purgatoire River and make the turn south, our usual route to the cabin. I'm watching for "the spot" and there it is. We pull over and before the car is fully stopped, my door is swinging open and I'm getting out. Above me is a vertical rock face, a deep layer of pale sandstone perhaps five feet thick that overhangs less solid rock like a guardian. This cliff seems like countless others around here, an exposed slice of geology, with grass and sagebrush sprouting in spotty clumps. But what is revealed at this particular cliff is momentous.

A cascade of coal debris and soil below the rock face makes a tricky scramble, but in minutes I am peering beneath the sandstone's brooding lip. Up close, the white line is fractured and uneven, no more than an inch high, but distinctive amid the other strata. I reach out a finger and touch it.

This unremarkable chalky line is the mark of an ancient cataclysm. Below it is a vanished world—the Mesozoic, the Age of Dinosaurs. I lay my hand flat against the crumbling coal, all that is left of that world. Then I lay my palm against the massive bulge of sandstone above the white line. Its surface is friction-y, fine grained. This is today—the Cenozoic, the Age of Mammals.

I'm a zoologist. I've always found geology interesting but so *passive*. Rocks don't sing in the spring or bugle in the fall. They don't fly or echolocate or climb trees without arms or legs. But they are storytellers, if we learn to read their language. To understand the significance of this white line, I become a short-term geology wonk.

On this cliff, Earth's past is exposed like a multilayer cake, each layer revealing a chapter in Earth's history. The icing between two of the layers—the K-Pg Boundary—holds the key to answering a great mystery.

It's actually what's missing above the boundary that makes it so significant. Below the white line, fossils of ancient plants and animals, particularly dinosaurs, are abundant. But above the white line, geologists find no evidence of dinosaurs, not a bone or tooth or claw or even a footprint, or evidence of almost any other animals or plants from the Cretaceous Period, the last period of the Mesozoic. Because after the thin white line, the dinosaurs were no more.

So what happened to the dinosaurs?

Sixty-six million years ago a massive asteroid collided with Earth off the coast of today's Yucatán Peninsula of Mexico. This is the event broadly accepted by science as ending the Age of Reptiles. The asteroid was so enormous (six miles across) and traveling so fast (47,000 miles per hour) that its impact bashed a crater twenty miles deep and one hundred miles wide. The debris and vapor from the impact volcanoed up, sending thousands of cubic miles worth of dirt, rock, vapor, and stuff into the atmosphere. Enough to fill the Grand Canyon, maybe two times over or more.

A cubic mile is a volume of debris one mile high, times one mile wide, times one mile long. The Grand Canyon is about a mile deep at its greatest depth, 277 miles long, and eighteen miles across. That works out to nearly five thousand cubic miles (if the Grand Canyon had straight sides and even depth for its entire length). Now imagine the Grand Canyon filled to the brim with pulverized rock, ash, and debris, and then all that stuff, or maybe twice that amount, propelled into the air.

A cascade of mind-boggling effects followed the impact—tsunamis, earthquakes, massive firestorms, and windstorms. All spreading around the globe.

These events destroyed countless animals and plants but would not alone have led to a mass extinction. It's what happened next that did. Those Grand Canyons of debris created a dark cloud of particles that dispersed around the globe, hanging in the atmosphere for years. This dark layer blotted out all sunlight, leaving Earth cold and pitch black. The world descended into an *impact winter*.

Unable to photosynthesize without sunlight, green plants died. Animals that ate green plants followed, as did those that ate the plant eaters. Entire ecosystems collapsed. The debris eventually settled, but carbon dioxide—CO_2—remained in the atmosphere, leading to a greenhouse effect, like what is heating up Earth today. Instead of cold and dark, Earth became an oven.

So what killed off the dinosaurs? Climate change.

In a short period of time, the asteroid impact led to a complete change of Earth's climate—first dark and cold, then blazing hot. Most things not killed by the initial cataclysm—with its fires, floods, and storms—died over the following months from cold, starvation, drought, or extreme heat. Up to 75 percent of all species on Earth—plant and animal—disappeared. It was a mass extinction, the fifth in the history of Earth. And it was caused by climate change.

Like an episode of *DinoCSI*, sophisticated science and geology detective work spearheaded by the University of California, Berkeley, geology professor Dr. Walter Alvarez figured out this ancient event.

The key to the mystery of what happened to the dinosaurs? The one-inch white line on the cliff face. This plain-Jane outcrop holds the physical evidence of the cataclysm that forever changed Earth and set the course for the world we live in today. It's clearly visible just three miles from our cabin and in very few other places around the globe.

The K-Pg Boundary is a layer of sediment that built up around the world after the asteroid impact particles settled out of the atmosphere. It contains iridium, an element found in asteroids and meteors but rarely on Earth. And shocked quartz, which results when ordinary quartz rock is deformed by a massive *pressure shock* and is commonly found in meteor-impact craters. And glassy microtektites, also from meteor impacts, which is where I'm going to stop cause this is way too wonky.

What isn't wonky is the immense tragedy of losing forever such amazing animals. Armor-plated triceratops placidly grazing on leafy greens to maintain their twenty-thousand-pound figures. Twelve-foot-tall T. rexes shaking the ground with their twelve-thousand-pound bulk of most-terrifying-predator-in-Earth's-history. Dinos with horns and spiked tails and wings and feathers. Dinos with duck bills and parrot beaks. Dinos that ran in packs and cared for their young. A magnificent carnival of animals and plants both vaguely familiar and delightfully bizarre. All wiped out after that terrible day, leaving no offspring, nothing to carry on their genes or tell their story except the remnants of their bones locked in ancient stone.

The term *K-Pg Boundary* deserves an explanation. On the geologic time scale, the K-Pg Boundary marks the transition between the Cretaceous Period (the last period of the Mesozoic Era) and the Paleogene (the first period of the Cenozoic Era). It's short for Cretaceous-Paleogene Boundary. *Pg* represents Paleogene, which makes sense. But why *K* for Cretaceous? Why not use *C*? Stay with me here. In an additional layer of konfusion, the geology community uses *K* as the abbreviation for Cretaceous because *C* was already taken for the Cambrian Period—which is much older (five hundred million years ago), and so like the oldest child, it gets priority.

But there's more. Cretaceous comes from *creta*, Latin for chalk, and *K* refers to the German word for chalk—*kreide*—which is also used for the Cretaceous Period. Perfektly klear! (I'm not sure why they used *Pg* for Paleogene but not *Cr* for Cretaceous.)

Don't unbuckle your seat belt yet. To make this all even more confusing, until recently the boundary was known as the K-T Boundary— that's how you'll often see it on maps and interpretive signs and in books and articles. *T* refers to the Tertiary Period, an outdated term no longer officially on the geologic time scale but still in broad popular use. The Period-Formerly-Known-as-Tertiary is now broken into the smaller time periods Paleogene and Neogene.

Clear as primordial sediment.

As a writer, I find the term K-Pg Boundary too cumbersome. If I were queen of the world . . . or at least the one responsible for naming geologic stuff . . . I think I could do better, for popular consumption anyway. The term is a mouthful and based on shorthand most people don't understand. OK for scientists but not really suitable for all audiences. We need a fun, tweet-ready name for this. K-Pig Boundary? ByeBye-Binosaur Boundary? Or just Dino-NoDino? Maybe the boundary just needs an image consultant.

───────────

CABIN JOURNAL—JUNE 2015: *The joy of watching storms from our cabin porch! I sit, coffee cup in hand, serenaded by a rumbling sky. Heavy clouds parade north to south in front of me, darkening the horizon. They slowly engulf Fisher's Peak, cloaking it deeper as rain advances across the ridges toward the cabin. A small patch of blue valiantly persists to the southeast, a window that is closing by the minute. Raindrops blow onto the porch as the storm arrives, tickling my arms with their light, cold touch. A nighthawk calls from overhead, though it is not yet 3:00 PM. The rain deepens and lightning flares—here, there, over there. The world around me darkens to a deep blue.*

Torrential rains flash down our arroyo, which begins as an ephemeral stream far back up on the flank of Montenegro. Countless rains over countless years carved it eastward across our land, cutting deeper and deeper until our humble arroyo earns a place on the topographic maps, and a name, Toro Canyon.

The rain is a sculptor's chisel, cutting into the landscape. Has it yet carved down sixty-six million years, into the deep past? Our land is less than three miles from the cliff face where I first scrambled up to the K-Pg Boundary and from another boundary site, in Trinidad Lake State Park near where Long Creek flows into the Purgatoire. As I increasingly see the impacts of present-day climate change on our land, which science is warning us may lead to Earth's sixth extinction, I learn this area is among a handful of places in the world where the evidence of the fifth extinction is easily visible. Is it exposed on our land too?

I go exploring. Toro Canyon is deep and narrow enough that in places I can touch both walls with my outstretched hands, the cutbanks rising above my head fifteen feet on either side. I stroke the raw tendrils of newly exposed roots, feel the rough topography of the rock. In places the pale sandstone overlays the strata below like a heavy brow ridge. The coal seam it shelters is iridescent in some light—a sheen of green, then of blue, like the plumage of a magpie. It shows many fracture lines, seeming fragile and crystalline, crumbling into sooty cubes at my touch.

I follow the streambed, which is dry much of the year but trickling now with the remnant of yesterday's rain. The streambed reserves enough water that lush grass thrives in shaded spots throughout the year and patches of wet sand hold the hoofprints of deer, the five-fingered pawprints of raccoons. The passage is tangled, crossed by fallen limbs or projecting switches of willow and three-leaf sumac. In many places broken slabs of sandstone make uneven steps. Where steep cutbanks keep out the sun the air is cool and damp, so different from the arid world of sagebrush and prairie lizards fifteen feet above my head. Walking Toro Canyon is an odyssey into hidden microclimates where sheltered alcoves fed by water seeping from the walls sustain tiny hanging gardens and patches of water-loving sedges persist in the dry streambed.

I peer beneath each of these alcoves, searching for the white line. There are no clear, uniform layers of rock here. In places the coal seam is prominent, in others it disappears. As the canyon gets closer to Long Creek, it widens and loses its character, spreading out into grassy meadows. If I am to find the boundary, it will be in the deeper recesses of Toro Canyon.

Through the summer, I explore the length of Toro Canyon, from our arroyo to the lower meadows, but I don't find the boundary. Rick nicknames me the Boundary Hunter. I am no geologist, just a curious amateur who is easily sidetracked by a bird or a plant or the tracks of a mountain lion leading up a branching arroyo. Still, the deep sandstone caprock and the coal seam are evident here in so many spots that the K-Pg must be sandwiched between them somewhere!

I keep searching through a circle of seasons. Is it here? I want it to be here. I've seen it just three miles away—I could walk to it! Straight down Toro Canyon, hang a left at Long Creek. But I want it to be *here*, on our property.

I know it's here. Buried, hidden. I'm so close. I'll keep peering in alcoves where bats roost, poking under the sandstone caprock and above the coal where pack rats nest. And one day, I'll find it.

————————

"There's another one." Olivia points to the impression of a leaf in the broken rock scattered along Elk Ridge. She kneels down and traces her finger along the lines of the leaf, the veins and edges as finely sketched on the sandstone as if a botanical illustrator had meticulously drawn them with pencil.

Then she points to an odd pointed spear shape with deep grooves and pits in no particular pattern. I stroke the impression with my fingertip. The pits and grooves seem much deeper than a leaf would make in wet sand. Whatever this is, it must have been harder and more three dimensional than a leaf. It almost looks like the ray of a starfish.

Winter is fossil season here, when the low-angle light casts the perfect shadows to highlight ancient plants captured in the sandstone. Fossils litter the landscape wherever rock is exposed or broken. There are long oval leaves with pointed tips, prominent veins, and smooth edges. Wide leaves with three pointed lobes that look vaguely like maple leaves. Those odd spears with deep pits and grooves. And countless large fan shapes of numerous blades radiating from a central base—instantly recognizable as the leaves of palm trees.

I find a 1976 paper from the New Mexico Geological Society—*Upper Cretaceous and Paleocene Floras of the Raton Basin, Colorado and New Mexico*—which has drawings. I make my way through its detailed discussion, which dates our fossils to the Paleocene Epoch, sixty-six to fifty-six million years ago. I take an amateur stab at identification. The oval leaves seem to be from ancient magnolia trees, the maple-leaf wannabes, the extinct ancestors of sycamores. The deeply pitted spears are probably corals (how cool is that?!), not starfish. And the palm fans? That one's easy. They're unmistakable, like avatars of palm trees I would see walking neighborhoods in Miami or Los Angeles.

We find no dinosaur bones because our fossils date from after the K-Pg cataclysm. I study the corals and the palm fans. The case they present is obvious. These fossils date from a time millions of years ago when this area was warm and humid, a land where palm trees grew on the edge of seas in which corals thrived. A world with a climate very different from today. These after-the-asteroid fossils are evidence of climate change. But there was no mass extinction that wiped them out, though they eventually disappeared, replaced by piñons and yucca and prickly pear cactus—plants definitely not from a humid landscape.

The key to this transition is "eventually." A changing climate led to the disappearance of palms and sycamores and corals, but over a great period of time. The change was gradual, at a pace that allowed plants and animals to adapt. The geologic record shows that Earth's climate has changed constantly over millions of years. Colorado has been steamy with rain forests, blown up by volcanoes, frozen beneath glaciers. It isn't the change in climate that upsets the apple cart, it's the pace at which it happens. And the current warming of the climate is happening very fast.

A primary indicator of climate change is the accumulation of CO_2 in the atmosphere. Too much of it and Earth becomes a giant, heated greenhouse.

It looks like the upraised tail of a rattlesnake. That's what I think when I see a graph of the growing amount of CO_2 in the atmosphere over the last three hundred years. When a rattler is threatened, it coils its body and raises its rattle high, like the abrupt spike of the graph. For the first two centuries, the graph is flat, the snake's body stretched calmly on the ground. Atmospheric CO_2 is a fairly constant 278 parts per million (ppm). Nothing to fear, no need to rattle.

In the late 1800s, the curve begins to creep upward. The engine of the Industrial Revolution is revving up, along with the burning of coal, oil, and gasoline. The snake is waking up, getting twitchy. But after 1960, the curve arcs dramatically skyward as burning of fossil fuels explodes to meet the needs of a booming human population, the growth of international trade, the rise of the Internet, and accelerating global consumerism. The rattler's tail is raised in alarm.

In May 2022, NOAA's Global Monitoring Laboratory measured more CO_2 in the atmosphere than at any time in at least four million years. Concentrations reached 417 ppm, 50 percent higher than in 1800, before the Industrial Revolution. Average global temperatures are 1.1 degrees Celsius higher. Climate change experts have long said we need to limit the average temperature rise to 1.5 degrees Celsius to avoid the worst effects.

The curve is rattling a critical warning. The pace of growth is what makes the rattlesnake curve dangerous. It's not as abrupt as an asteroid collision, but it's a pace at which most species cannot adapt quickly enough to survive. It's a pace that could lead to the sixth mass extinction of life on Earth.

Human action is the cause of the pace. We are the asteroid.

———————

Geologists tell us there have been five mass extinctions in Earth's history, the K-Pg event being the most recent. By intense study of the fossil record, scientists can theorize the likely causes of events multimillions of years in the past. (Of course, they don't all agree!)

The first mass extinction, known as the Ordovician-Silurian Extinction, happened 440 million years ago. Eighty-five percent of species went extinct.

The cause? Climate change. Global temperatures plunged, glaciers formed, and sea levels dropped. There was no life on land yet, and most marine life could not survive the cooling and shrinking of the oceans.

The Devonian Extinction was the second. This period is known as the Age of Fishes, 365 million years ago.

The culprit? Climate change. This one actually happened over millions of years and comprised a series of pulses in the climate—global warming, then cooling; sea level rise and fall; reduced oxygen and CO_2 in the atmosphere—that led to extinction of 75 percent of species.

Number three was the Permian Extinction 253 million years ago, known as the Great Dying because it nearly snuffed out all life on Earth.

Reason? Climate change. Scientists debate the cause of the changes: either intense volcanic activity spewed toxic gases into the atmosphere, acidified the oceans, wrecked the ozone layer, and created a greenhouse warming effect; or an asteroid may have struck Earth, raised a pall of dust that blocked the sun, and caused acid rain. Whatever the trigger, massive changes in the climate led to destruction of 96 percent of sea life and 70 percent of life on land.

The fourth happened about fifty million years after the Great Dying. The Triassic-Jurassic Extinction, 201 million years ago, led to loss of 80 percent of species.

The guilty party? Climate change. Volcanic eruptions sent huge amounts of carbon dioxide into the atmosphere, heating the globe, melting ice sheets, and raising sea levels.

Global climate change likely caused all five of Earth's mass extinctions. Not surprising, really, because a global extinction event needs a cause with global reach.

The similarity between these past extinctions and what's happening today is the cause: climate change. The difference is that no species living in those eras had the ability to either cause or alter what happened.

We do.

I appreciate how Dr. Alvarez, in his book *T. rex and the Crater of Doom*, animates for us the world on the day the asteroid hit. It was so very long ago, beyond our mind's ability to truly grasp. We imagine the world of the dinosaurs as an artist's rendering, its animals reconstructed skeletons hanging in museums. But it was a real world, a lush and vibrant world, populated by complex, diverse life-forms. It was as real as Elk Meadow and our pine forest; its grazing hadrosaurs and hunting T. rexes and soaring pteranodons as real as our elk and bears and circling ravens.

It was a terrible day, the last day of the world that was. It makes me sad that dinosaurs, such wondrous animals, never graced Earth again and never will. Countless other fascinating species have disappeared through eons of time, but not so abruptly or totally. I will never see cave bears or dire wolves, mammoths or giant bison, but I have seen their descendants. Dinosaurs, however, are gone completely.

In a strange, big-picture way, I find hope and solace that some plants and animals have survived every mass extinction and life always resurged. New wonders evolved to fill the many niches of a brand-new world, as splendid in their way as those that are no more.

Survivors of the asteroid included distant relatives of dinos—turtles and snakes, frogs and crocodiles—and the smallest of the feathered dinosaurs that would become birds. And small, warm, furry creatures that could shelter beneath the destruction and feed on the detritus. Those primitive mammals became, eventually, us.

It's worth a moment to consider what the world would be today if the asteroid had missed, if its trajectory had been off, if it had passed Earth by instead of colliding with it. The late Cretaceous was a fairly stable world. Dinosaurs had thundered, plodded, and scampered on Earth for 150 million years as the dominant animals and might have remained the headliners for hundreds of millions more, with mammals relegated to the minor role of small scurriers. But it was not to be.

The asteroid hit, obliterating the dinosaurs and ending the Age of Reptiles. And, it's worth noting, also obliterating the asteroid. It took millions of years for Earth to recover and new species to evolve and flourish. But they did. Eventually primates emerged, and eventually some of them abandoned life on four legs to stand upright. They were relatively small and weak, lacking in sharp teeth or claws, unable to fly and not particularly fast. Their young were born helpless and undeveloped, like the hatchlings of bluebirds, needing years of parental care. The only gift they had was a large brain. But that was enough.

Sixty-six million years after the dinosaurs, the descendants of the small scurriers are the dominant life-form on Earth. In the way of the world, we are the beneficiaries of the asteroid Armageddon and the fifth extinction—they made way for our ascendance. Another sixty-six million years in the future, Earth's biota will likely be very different, whether or not we avoid a human-caused Armageddon in our time. Life will journey onward. All the familiar species we love—bluebirds and columbines, honeybees and chorus frogs, mountain lions and tiger moths and bears—might have vanished. *Homo sapiens* may likely be gone or so changed we would not recognize our descendants. But life will persist and new wonders will take the place of old. It offers little reassurance at a personal level, but I take comfort in it. And it leads me to ask a question: What mark in the geologic record will our era leave?

16

BLUEBIRD LESSONS

Probably the question I am asked more often than any
other is: Do you honestly believe there is hope for our
world? For the future of our children and grandchildren?
And I am able to answer truthfully, yes.
—*The Book of Hope: A Survival Guide for Trying Times*,
   by Jane Goodall and Douglas Abrams, 2021

We are on a dangerous path, but we have the power to chart
a better one. Still, change will come only if we demand
action from the public officials who represent us and the
businesses we support.
—*Survival by Degrees: 389 Bird Species on the Brink*,
   National Audubon Society report, October 2019

May 2022: It's a morning in early May and we're on our way to
the cabin, driving south along Interstate 25. Along Plum Creek,
south of Denver, the wild plums bloom like white clouds snagged among
the bare-branched willows and chokecherries. Despite a slight haze,
the sky is blue and the sun shining. We drive over the seventy-three-
hundred-foot summit of Monument Hill and begin the descent into
the Arkansas River Valley. Ahead, a heavy pall clouds the air.

As we drop down into Colorado Springs, the wind thrashes the
trees and runners along the trail through the Air Force Academy bend
their heads to the wind. A brown mist moves across the city from the

south, engulfing its buildings and cloaking Pikes Peak until only the snowy summit glows through the shroud.

The haze is a mix of dust blown by the fierce winds and smoke coming from New Mexico, where a wildfire has burned across two hundred thousand acres and would soon become the largest wildfire in New Mexico history. The fire is one hundred miles away but being driven by strong winds. Like so many wildfires in the West today, it is born of the usual climate change suspects: extremely dry conditions and above-average temperatures.

South of the Springs we pass a smoldering black patch of grass between the highway and the frontage road, maybe thirty feet from our car. A cloud of smoke streams across the highway and we drive through it, smelling the acrid scent of burning hay. As we start to call 911, we pass another small grass fire, then another. A pickup is pulled over by the first, and a guy in work boots is stamping on the blackened grass. A yellow-vested highway worker is shoveling dirt on the second. Then we see Colorado Department of Transportation crews working on two more grass fires farther south.

We pass through Pueblo, its air thick with the brown haze. We can't see the Wet Mountains, which rise just to the west of the highway. Only the tip of Greenhorn Mountain pokes above the brown, a sliver of white snow glowing at its summit.

We come up a rise and the landscape opens in a broad expanse of winter-dead range where the High Plains meet the mountains. We're more exposed now and the winds buffet us, pushing our Toyota Highlander around like a bully shoving the new kid. There is a semitruck blown over on its side in the median.

Beyond the blown-over semi, a couple dozen more semis, RVs, and other high-profile vehicles are pulled over, not taking the risk of blowing over themselves. They could be stopped here for hours; the winds aren't likely to die down until evening.

We pass another semi blown over in the northbound lane and cautiously pass a towering passenger bus, wondering why it is still underway with a cargo a lot more fragile than other freight. Ahead of us, a large pickup towing an open trailer loaded with rolls of grass sod begins to fishtail, its heavy trailer swinging wider and wider in a pendulum that threatens to swing the truck off the road. We make it past the truck but start to worry—what will the cabin be like today?

But when we reach Trinidad, the air patterns favor us. The wind is calmer, the sky a magnificent blue. Fishers Peak stands clear and sharp atop Raton Mesa, and to the northwest, the Spanish Peaks rise in crystal beauty. But the wind still gusts with vigor, flexing its muscles.

At the cabin, our meadows are dry, but beneath the tufts of last year's grass, green sprouts push up. Sand lilies lie among the brown like white stars fallen from the sky in the night. Tiny white asters, velvet tufts of sagebrush, various yellow composites all poke up, announcing spring.

I fill a hummingbird feeder and hang it on the deck. Will they come? For an hour or more, I hear no buzz of rapid wings. Finally, a male black-chinned hummer drones in to feed.

By 3:00 PM a half moon rises over Raton Mesa, a slice of lemon poised against the blue. The moon, at least, has not blown away in the wind.

On the power line along the drive, three western bluebirds perch. One leaves the wire and flies to the piñon across from the deck. The tree that shelters the cabin nest box. I see a flash of red breast plumage in the sunshine: a male.

He lands on a branch of the tree and pauses, surveying. Then he hops to a branch closer to the box, and before I can whisper, *It'll make a great home*, he darts through the hole.

The next day we make a tour of the nest boxes, taking inventory. Carefully I open the cabin box. A grassy nest holds five perfect blue eggs.

In spite of fire and wind and drought and so many questions about the future, the bluebirds have returned. The next bluebird season has begun.

———

I didn't set out to become a witness to climate change. But if I had, I couldn't have chosen a piece of land that better represents the many chapters of the story. Our land tells the saga of climate change in microcosm—a century of fossil fuel extraction; extremes of weather; drought; species change; wildfires; even the mark of the last mass extinction, exposed here like a warning. This landscape that shows so many impacts of climate change was also a party to creating it. Like many unwitting chroniclers, we unknowingly captured the evidence of all this in our journals. Our sightings are the raw data of climate change. The hundreds of entries about birds and nestlings, what flowers are blooming, the movements of bears and elk, the weather, reveal changes that are only evident over time.

I began this book with the idea that we have broken our covenant with the natural world, leading to the current climate crisis. The Dust Bowl of the 1930s resulted from a broken covenant with the land of the Great Plains. Through diligent stewardship, that covenant was restored and the land has healed. So, too, can we repair the broken trust that has led to a changed climate.

If I am a witness to change, I am also a witness to lessons that can help us face it, for in our journal are countless stories of recovery and renewal. This book might seem to be a story about loss, about a world teetering on an abyss of tragic change. But it is actually a story of hope. My piece of the wild may be under grave threat, but it is still a place full of wonders, of joy and renewal. The bluebirds are back, carrying on in the process of carrying on. New growth is coming up through the dead and dry. In all of this there is hope.

My observations of bluebirds and other species have taught me about resilience. About recovery and grit. About viewing life over many seasons, not just one. That embedded in all of us—bird, mammal, plant, and everything else—is the will to live. Is hope.

Life wants to *live*. Living things strive and fight and persist. Chorus frogs and Woodhouse's toads spend long months entombed in the dry bed of the pond, waiting. They wait—for rain, for release, for the chance to procreate—because within them is hope.

Blue grama and sand lilies lie hidden through long winter months, sheltering out of sight within the vitality of their roots. Spring will come, their essence tells them, the rains and days of long sunshine will come. And they have hope.

Bluebirds who lost all their nestlings to drought or bears the previous year, who spent a full season of their short lives striving to rear young ones only to fail, still return the next season to nest again. Because within them there is hope.

These are not the foolish hopes of a gambler betting endlessly, sure their luck will turn. This is the pile-driving, embedded-in-DNA impetus to survive, and it cannot be denied.

Nature heals, given any chance at all. Life on Earth recovers. But if climate change continues unabated, would that recovery be in a time frame that includes us? If not, we may share the fate of the dinosaurs.

There is a lesson for us in the K-Pg Boundary. The asteroid hit and changed the world instantly. Dinosaurs and the creatures of the Cretaceous had no time to react. We do.

Dinosaurs lacked the ability to change their environment. Their only strategy was physical adaptation, but that takes time, which they didn't have. Humans have an astounding ability to change our environment—it's what got us into this mess to begin with. We are capable of amazing things, especially when we work together. We are the driving force behind global climate change, but we are also the answer to mitigating it. What we have wrought, we can unwrought. At least enough to avert the worst of the consequences.

This landscape has repeatedly recovered from environmental damage. One hundred years ago, coal mining and coking left the hills bare of trees and the air dark and noxious; now, piñon-juniper forest again covers the land, and the air is so clear that on moonless nights the stars are diamond-sharp. Shrubs and trees erase old roads cut during ranching and mining days. New piñons and junipers reclaim meadows cleared in past decades to create grazing land.

In the time frame of life on Earth, modern humans are infants. *Homo sapiens* have been around only some 250,000, maybe 300,000, years. But we are some pretty powerful infants. We've altered the world and impacted environmental systems so profoundly that we have likely put Earth on a trajectory that will wipe out much of organic life. That's immense power and ability for such a young species. And in that power lies our hope.

Over thousands of years, we've applied our abilities to alter the world to benefit ourselves, which is what we are biologically driven to do, and this made our species more successful, with higher survivability—in the short term. In the long term there is a tremendous price for these alterations. Altering Earth for humans has put us on a runaway train that will end in a devastating crash, unless we act. It's beyond time to talk about it. Time to *do* about it.

A couple of first steps on that road.

Stop doubting the science. The data on the changing climate is so overwhelming, from so many decades and disciplines, that it's incontrovertible. The dodges and dismissals from those who want to continue in denial are getting weaker. The United Nations' International Panel on Climate Change (IPCC) has massive amounts of constantly updated data. Of course, doubters gonna doubt, regardless of overwhelming evidence, especially if there's a profit motive. So it's up to the rest of us.

We must get beyond the idea that climate change is primarily a political issue. It's an environmental, health, economic, and sociological issue that's been made into a political issue for the short-term benefit of some interests. When I told a friend I was writing this book, she commented that I needed to look at all sides of the issue. "What sides?" I asked. It's not an opinion; it's science. How many people, if diagnosed with cancer, would stop to debate the issue? They might research best treatments but would accept the science and go all out to battle the disease.

Climate change is not in the future, it's now, and it's threatening all of us. When a house burns in your neighborhood, you don't debate whether it's a hoax or whether it's politically expedient to battle it. The fire department attacks the fire full-on, with everyone's support. We must do this with climate change because our global house is on fire and we're all in imminent danger of burning up.

We need the political will to fight climate change. We have the tools and technology to tackle the climate crisis, we just need the societal commitment to go for it full-on. According to the IPCC, the cost of solar, wind, and batteries has dropped as much as 85 percent. Renewable energy is often cheaper now than fossil fuels. "The barriers are entirely political at this point," wrote the *Los Angeles Times* editorial board in an April 5, 2022, op-ed, "maintained by politicians and the fossil fuel interests they prop up."

Don't fall prey to pessimism and apathy. When the bluebirds lose a nest of babies, they don't give in to pessimism. They nest again. Stories of human resilience and triumph in the face of disaster are beyond counting, and they teach us that we can and must fight this crisis. Having researched, written about, and been involved in conservation and environmental issues since I was a teenager, I know just how easy it is to fall victim to pessimism. But there is really no point.

Still, there is an awful lot of grim environmental news, so how do we react? Another friend told me it was too late, that greenhouse-gas-caused warming could not be reversed. She was just going to make things the best possible for herself and her family, find a patch of land somewhere, and go into survival mode. A bit dramatic, I thought. A bunker mentality isn't helpful to anybody but those selling bunkers. And where would she go, anyway? We can't all head north to Canada and Alaska, all 330 million of us in the United States, not to mention the millions of climate refugees already fleeing worse-off areas. Canada would build their own border wall to keep us out.

And what about our piece of the wild? If it's all going to end as *Mad Max: Fury Road*, are we giving up now? Have we abandoned putting up nest boxes because the temperature models are so worrisome? Of course not. We aren't giving up and we aren't pretending nothing is changing. We will continue to appreciate and celebrate nature and work to preserve the species on our land. And advocate for action *now* on climate change. There is so much to do, we must all embrace it and get going, work to correct this unintended debacle of our species' own making.

And that hope thing? Here's where I see the rays shining through.

Economics will drive change. Humans are notoriously bad about doing the right thing for the right reasons. It's biology. We're programmed to maximize our own outcome, meaning we're really good at doing the right thing for personal benefit, including financially. So economics will be, and already is, a primary driver for climate mitigation. I think it is what will carry the effort to overcome the climate crisis.

The great news is that the future of the green economy based on low-carbon, renewable, energy-efficient, environmentally friendly, and socially inclusive principles and policies is bright (green). A 2019 analysis by University College of London geographers found that the green economy employed an estimated 9.5 million workers in the United States (4 percent of working-age people), is increasing by more than $60 billion per year, and would reach $1.3 trillion as a sector of the economy.

Jobs will be lost in the fossil fuel sector, but many more will be created in renewable energy and other green industries. The International Labour Organization projected the green economy could create twenty-four million new jobs worldwide by 2030. There is enormous opportunity in the dawn of this new age, and savvy business people and companies—from entrepreneurs to fossil fuel industry giants like Exxon—are already planning on and investing in it.

Youth is our hope. Like bluebird nestlings, our young carry the future, but not just as the bearers of our DNA. Unlike bluebirds, our nestlings aren't limited to blindly following instinct. They have conscious hopes for the future and the ability, energy, and passion to affect that future. Worldwide, many young people are determined to protect their future. Our young are not birds. They can make choices, seek solutions, take action. They already are.

Swedish climate campaigner Greta Thunberg exemplifies the next generation's demand for climate action. With the brutal honesty of youth, she lambastes politicians and leaders worldwide for stealing her future and angrily demands action. Dozens of youth-organized and youth-led groups and coalitions, including Youth Climate Movement and Fridays for Future (gotta love one of their slogans: "There Is No Planet B"), are marching, lobbying, advocating, and demanding climate action of political and business leaders. It reminds me of the youth-led anti–Vietnam War movement of the 1960s and the actions of young people during the civil rights movement who registered voters, sat at lunch counters, and protested, in spite of beatings, arrests, and worse.

But youth can't make effective climate change action happen by themselves. Every generation holds the world in trust for the next and has an inherent covenant with their descendants to be good stewards of that world. We in older generations haven't done the best job of that, and now we can't just sit back with a martini and leave it to the kids. We must work with younger generations to change the climate trajectory.

Olivia is in her midtwenties now. She and her generation will write the next chapters of *Bluebird Seasons*. She lives on the West Coast and loves the big city, but her childhood among wild things at the cabin showed her the importance of a healthy natural world. With their lives so dominated by social media and the Internet, she says, her generation really cares about nature and sees its value. For them, nature is an antidote to the overwhelming virtual life they must live. "We all hate social media," she says. "We know it's shallow, but we have to use it.

If you're not, you've removed yourself from society in a big way. But nature is real. It is what it is." Her generation will work for and demand solutions to climate change. "What the planet gives us—fresh drinking water, food, clean air—is what we all *really* need. Social media means nothing if we're all dead."

What does she think of the future, in light of climate change? "I'm hopeful, but my generation is definitely angry. We're the ones who are going to have the impact of climate change, and younger generations even more. We care about this issue, and we want change." Clean energy, sustainability, preserving wild places—her generation is committed to these things. Not everyone, of course. No generation is homogenous in its attitudes.

"My generation will demand that things be different, we'll work for it. The collective sentiment has shifted to being about conservation. Nature is a motivator for us; we see hope in it. We are a very moral generation, more and more people are insisting on sustainability, social justice, a healthy planet over profits. In a capitalist society, collective consciousness weighs heavy, and business has to meet our needs and demands. We're the customers."

Many of her friends are actively committing to sustainability in their lives and work. Anna Rose, Olivia's best friend since babyhood, who has visited the cabin many times, is a fashion designer in New York. She's working toward a career in sustainable fashion—clothing that is designed, manufactured, and distributed in ways that are low polluting, conserve water, have a small carbon footprint, and are socially equitable. Fashion, she tells me, is the number-two most polluting industry in the world, after fossil fuels. Fast fashion—cheap, trendy styles meant to be worn a few times and discarded—is hugely wasteful, polluting, and often based on poor labor practices, including the use of child and forced labor.

Many in her generation reject fast fashion. There is a big rise among young people to buy used fashions through thrift shopping or online secondhand sites.

"I'm optimistic, but it's hard to be all the time. Sometimes things look bad, but overall I'm hopeful. At the end of the day, I think people are generally good and sustainability needs to happen, so I hope it will."

CABIN JOURNAL—JUNE 2022: *I carefully open the side of the cabin nest box. Numerous bright black eyes gleam back at me. The box is filled with feathered bodies—fully fledged young bluebirds magically transformed from the five perfect blue eggs this box held six weeks ago. Soon they will launch from this natal box into the world to grow and thrive and, with luck, return here next year to raise their own babies.*

Emily Dickinson wrote, "Hope is the thing with feathers / That perches in the soul / And sings the tune without the words / And never stops at all."

In spite of so much that seems hopeless, hope never stops singing.

The young of the latest bluebird season are bright-eyed, undaunted, quivering with life. They are the things with feathers that perch in my soul. Because of them and all the hope inherent in the wondrous world around me—in youth, in people, in Earth's resilience—I keep my eyes looking forward, toward endless bluebird seasons to come.

ABOUT THE AUTHOR

Richard K. Young

Award-winning writer, naturalist, and zoologist Mary Taylor Young has been writing about the landscape and heritage of Colorado and the American West for thirty-five years. Her twenty-two books include *Land of Grass and Sky: A Naturalist's Prairie Journey* and *Rocky Mountain National Park: The First 100 Years*. She received the 2020 Lifetime Achievement Award from the Colorado Authors League, was inducted into the Colorado Authors Hall of Fame in 2019, and was the 2018 Frank Waters Award honoree for exemplary literary achievement and a canon of writing that communicates a deep understanding, celebration, and love of the West.